AF797093

Science Fair Projects

BY

JOHN W. RUSHIN, D.A.

COPYRIGHT © 1999 Mark Twain Media, Inc.

ISBN 1–58037–100–0

Printing No. CD–1327

Mark Twain Media, Inc., Publishers
Distributed by Carson-Dellosa Publishing Company, Inc.

The purchase of this book entitles the buyer to reproduce the student pages for classroom use only. Other permissions may be obtained by writing Mark Twain Media, Inc., Publishers.

All rights reserved. Printed in the United States of America.

TABLE OF CONTENTS

© Mark Twain Media, Inc., Publishers

THE EFFECT OF SALT ON THE GROWTH OF BEAN PLANTS
SALT
SUDSY SOAPS
ACID RAIN: HOW IT AFFECTS PLANT GROWTH

INTRODUCTION FOR THE PARENT OR TEACHER

This book was written primarily to serve as a helpful guide for middle grade science fair projects. All of the discussions, explanations, and activities in the sections titled Introduction, The Big Question, Choosing a Topic, Finding What Is Known About My Topic, Making a Hypothesis and Designing an Experiment, Making Sense Out of My Project Results, My Conclusions, Sharing My Results, and Science Investigations Review Games are written for student use.

The sections titled A Time Line and Some Standard Letters and Forms, 22 Teacher "Do's," and Some Parent "Do's" deal with the organization and guidelines of events such as school science fairs and science academies. Therefore, these sections are of more interest to middle grade science teachers and parents than to the students. It is helpful for a science teacher (or parent) to read these sections before starting his or her student(s) out in the first section of this book.

INTRODUCTION FOR THE STUDENT

What is science? Why is science important? These two questions need to be considered by anyone who explores the world through a science investigation. One way that **science** has been defined is as **a way of solving certain types of problems that deal with the material world**. Experience shows that the best way to really understand the true nature of science is to become involved in a science investigation from start to finish. This book will guide you through that process.

All of our scientific knowledge comes from careful observations of the **material** or **physical world** in which we live. (The material world is defined as all of the matter and energy that we can observe and, sometimes, manipulate.) Scientific observations are typically used in three distinct types of research studies: (1) A **descriptive study** describes the way things are observed to be without trying to explain their causes. (2) A **correlation study** is really a type of descriptive study that attempts to show connections or associations among separate observations without establishing any causes. (3) An **experimental study** manipulates all possible factors, or variables, that might change in an experiment in order to establish the cause of observations. This book will provide opportunities to work with all three types of research studies, but experimental studies will be stressed.

Scientists often use what is known as the **scientific method** in their research studies. This method usually involves the following steps: (1) asking a question, (2) reviewing what is known about the question, (3) formulating a hypothesis, (4) designing a study to test the hypothesis, (5) observing and organizing the results of the study, (6) analyzing the results and drawing conclusions, and (7) sharing the results of the study with others. It should be noted that scientists do not always precisely follow every step of the scientific method. However, most scientists generally agree that the steps of the scientific method provide an excellent **framework for organizing scientific studies**. This book will explore all of the steps of the scientific method and will provide you with ideas and activities that show how these steps can be used in science investigations.

Science cannot answer all questions. In fact, science cannot answer any question with absolute certainty, because we can never really know what will be learned from future studies. However, science can provide very strong support for ideas and theories about the material world. The knowledge gained from scientific studies may also lead to technological advancements that improve our lifestyle. Think of the many ways that electricity can be used due to our basic scientific understanding of its principles. Or what about the many lives that have been saved by vaccines for diseases such as polio, smallpox, or diphtheria, which were all based on our scientific understanding of the human body and the germ theory of diseases? The benefits gained from scientific advancement may be reason enough to learn about science. However, the benefits of doing a science investigation also include the development of certain higher-order process skills, such as observing, measuring, classifying, inferring, predicting, formulating hypotheses, experimenting, interpreting data, and communicating. An ancient Chinese proverb says, "I hear and I forget, I see and I remember, I do and I understand." The information and activities in this book are designed not only to help you to understand and do science, but also to help you enjoy the challenge of designing, completing, and sharing your very own science investigation.

© Mark Twain Media, Inc., Publishers

THE BIG QUESTION

Good and Bad Questions for Science Investigations

Science uses observations and measurements to answer many questions about the material world, but not every question can be answered by science. Read the following list of 10 sample questions.

1. What causes a plant to grow?
2. Which super hero has the most power?
3. Does the color of light affect the growth of sunflower plants?
4. How many guardian angels can a person have?
5. Can ultraviolet radiation be harmful to humans?
6. Will sunflower plants grow better in plastic or clay pots?
7. Which good luck charm is the luckiest?
8. Does boiling geranium seeds affect their seed germination?
9. How do bees fly?
10. Will a strong hurricane hit the coast of Louisiana in the year 2025?

A good science investigation usually involves a question that can be answered by selecting one answer from a few possible answers. Another characteristic of a good science investigation question is that it involves a topic about which you can find a lot of background information in books, magazines, encyclopedias, and other sources. It should also be about something that is easily observed and measured. Finally, a good science investigation question should lead to an experiment that you can perform with available equipment and other materials.

Science usually does not deal with things that cannot be observed or measured. Things such as supernatural beings, forces, and events are generally not considered scientific for this reason.

Name ______________________________ Date ______________

Student Exercise

1. Which of the questions on page 2 can be answered by a science investigation? (List the question numbers.)

____

2. Of the questions that can be answered by scientific investigation, which ONE would be easiest for you to investigate as a science project?

____

3. Why did you choose the question you selected above?

____

____

4. Does the question you have selected meet the following criteria? (Circle yes or no.)

   (1) can be answered by selecting one answer from a few possible answers
   (2) is a topic about which you can find much information
   (3) is about something that is easily observed and measured
   (4) involves an investigation that can be performed with easily obtainable equipment

   Yes No

5. If the question you have selected does not have all of the above characteristics, list the criteria that are lacking.

____

____

____

6. If the question you have selected does not have all of the above characteristics, select another question from the list that does and write it below. Leave the following space blank if the question you have selected has all of the characteristics.

____

____

____

7. Which of the list of 10 sample questions on page 2 cannot be answered by scientific investigation? (List the question numbers.)

____

The Best Questions for Science Investigations

In review, a good science investigation question (1) can be answered by selecting one answer from a few possible answers, (2) involves a topic about which you can find a lot of background information, (3) is about something that is easily observed and measured, and (4) leads to an investigation that you can perform with available equipment and other materials.

You are now well on your way to understanding **The Big Question**—for your science investigation project, that is.

A question such as "Does Scarring the Seed Coats of Honey Locust Seeds Affect Their Germination Rate?" can be considered a good science investigation question. For one thing, it will force you to make a choice from a short list of possibilities. Nicking the seed coat will either increase seed germination, decrease seed germination, or will have no effect on seed germination. (You will learn how to design and carry out an experiment to test this type of question later). Another reason why this is a good science investigation question is because researchers have done a lot of work on seed germination, and there are many articles and books about this topic available in the library. Also, whether seeds germinate or not can be easily observed and counted. (Information about tables, graphs, and averages will come later). Finally, the materials needed for nicking and germinating the seeds are easily obtainable.

Of course, the **very best type of question** to investigate in a science project is **one that meets all of the characteristics described above and that has never been done before**. Although it may seem highly unlikely that a middle school student could discover information that is new to science, you should keep in mind that there are many good science investigation questions that have never been attempted. Many books describe the precise requirements for germination of geranium, pale purple coneflower, Kentucky coffee tree, and other plant seeds. However, the precise germination requirements for some plants have not yet been determined and described. After searching through books and articles for what is known about honey locust, you may find that its precise seed germination requirements have not been worked out.

© Mark Twain Media, Inc., Publishers

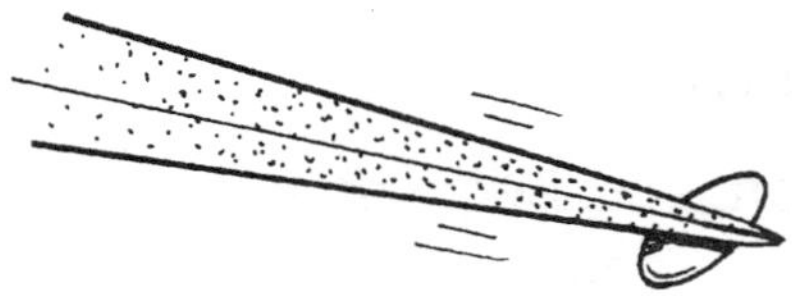

Even if the effect of nicking the seed coat of honey locust has been determined and described, your question might be changed to include other factors or combinations of factors. A search of the books and articles about seed germination should reveal that many seeds require nicking (officially called **scarring**) before they will germinate. This results in a large number of unscarred seeds that will form a seed bank for the production of future trees. The fact that acid from the stomachs of animals that eat certain fruits and seeds can increase the germination rates of seeds that have passed through the digestive systems of these animals is also known. Another good but more complex science investigation question would be to ask about the combined effects of scarring and acid on honey locust seed germination. It is less likely that the combined effects of these two factors (scarring and acid) have been worked out, so your question might be considered **original** and **creative**. (A word of caution: You must be very careful not to pick a question that is too complicated and too difficult to answer by a single science investigation.)

CHOOSING A TOPIC

Students are often glad to do science investigation projects because they are tired of **"learning about"** facts and concepts and look forward to actually **"doing"** science. Completing a science investigation project can be a very exciting and enjoyable experience if you pick a topic that interests you! Since selecting the right topic is so important to the success of the project, you should explore your science interests before deciding on a topic to investigate.

Things That I Wonder About (Science Interest Survey)

In today's world, science is a part of most of the things we experience. In sports, you may use scientific methods to compare the action of two different golf balls in a physics experiment. In a biology experiment, you might compare the effect of different water temperatures on the blood pressure of swimmers immediately following a race. Almost any topic that you can think of uses at least some science.

Science is sometimes divided into two main areas: the physical and the life sciences. The **physical sciences** concentrate on nonliving matter and involve subjects such as physics, chemistry, geology, meteorology, and astronomy. As you probably know, the **life sciences** (also known as **biology**) concentrate on all kinds of organisms, such as bacteria, molds, worms, plants, sponges, fish, birds, and humans. Life science subjects include molecular biology, microbiology, botany (plants), zoology (animals), ecology, and animal behavior. The subject of psychology may also be thought of as a life science, since it deals with the living organism you are most interested in—man. The subjects under the physical and life sciences make up the **basic science categories**. These science categories are a good place to begin your science interest survey.

Name ______________________________ Date ______________

Student Exercise

1. Write down two things from the physical sciences that you are interested in but cannot fully explain.

   a. ____

   ____

   b. ____

   ____

2. What basic physical science categories are your two answers for question one included within? (Circle one.)

   a. physics chemistry geology
      meteorology astronomy other:

   b. physics chemistry geology
      meteorology astronomy other:

3. Write two things from the life sciences that you are interested in but cannot fully explain.

   a. ____

   ____

   b. ____

   ____

4. What basic life science categories are your answers for question three included within? (Circle one.)

   a. molecular biology microbiology botany zoology
      medicine and health ecology animal behavior
      psychology other:

   b. molecular biology microbiology botany zoology
      medicine and health ecology animal behavior
      psychology other:

© Mark Twain Media, Inc., Publishers

Name ______________________________ Date ______________

5. Are you more interested in the physical or life sciences? Why?

6. If you had to choose one basic science category as your favorite, which one would it be?

7. Can you think of a recent television show or movie that deals with your basic science category? Name and describe this show or movie.

8. Name two professions or jobs that deal with your favorite basic science category.

© Mark Twain Media, Inc., Publishers

Narrowing My Topic to a Question

A student working in this book (let's call her Mary) wondered, "How do giant trees of so many different types grow from small seeds?". Mary wrote this question down as one of the answers in question three on page seven. This question definitely fits into the botany category, because trees are plants. However, **botany** is a very broad category that covers all plants. If Mary is primarily interested in trees, she might also explore the more specialized category of **forestry** in order to find more information. She should also communicate with professional foresters from the U.S. Forestry Service, the State Department of Conservation or Natural Resources, or a private forester. Besides providing valuable information about her topic, professional foresters can also provide insight and motivation about forestry as a career. In case you have not heard, **many science careers begin with student science investigation projects**.

The question of **"How Do Giant Trees of So Many Different Types Grow from Small Seeds?"** is a scientific question because it deals with material and measurable things—trees and seeds. There is also a great deal of background material in the library on both trees and seeds. However, some of the criteria for a good science investigation question (page four) are not met. The major problem with this question is that it cannot be answered by **selecting one answer from a few possible answers**. There is an extremely large number of different kinds of species of trees, and many factors are involved in the growth of trees from seeds. It would be highly unlikely that such a broad question could be answered by a single science investigation project. It is okay to think BIG at the start of your project and to come up with interesting questions like this one, but eventually you must **narrow your question to a manageable topic—a question that can be answered by a single science investigation project**.

As Mary reads more about her topic, she will find that the thorny honey locust tree is a very important tree in the early stages of some forests, where they spread from underground stems (called **suckers**). The honey locust tree also produces long, twisted pods (fruits) with seeds inside. These pods and seeds are eaten by animals such as deer and herbivores (plant eaters). The seeds may be "nicked" by the chewing action of the herbivore and then treated with the acid from the animal's stomach before being deposited on the ground with the animal's waste. Mary may find that the honey locust seeds can last for many years in the soil before they finally germinate and grow into a tree. She may not find exactly what factor triggers germination in honey locust seeds.

After reading some books and articles about the general topic of seed germination, Mary may decide to narrow her science investigation topic question to **"What Effect Does Nicking the Seed Coat Have On the Germination Rate of Honey Locust Seeds?"**. (Based on the information that she gathered from her reading, Mary could have also decided to ask a question about the effect of acid treatment on germination, or, in a more complicated experiment, the combined effects of both nicking and acid treatment.) This is a good science investigation question for several reasons that are described on page four, but especially because **it is narrow enough to be answered by selecting one answer from a few possible answers**. This question will also lead to an investigation that Mary can perform with easily obtainable equipment and other materials.

Student Exercise

A student named Ted wondered about and became interested in **animal cloning** and the possibility of human cloning. In Ted's search through books and articles about cloning, he read about a tiny aquatic flatworm called a **planarian** that can be cut into several pieces, each of which grows back into a clone of the original planarian. (Another term for this process is called **vegetative reproduction**.) In fact, during certain conditions in a stream, planaria actually break into smaller pieces and clone naturally. These little flatworms are easy to find in ponds and streams, but they can also be purchased from almost any biological supply company. (A list of some of these companies is provided later.) Planaria have been used in numerous regeneration and vegetative reproduction experiments by scientists for many years, and there is much information about them in books and articles.

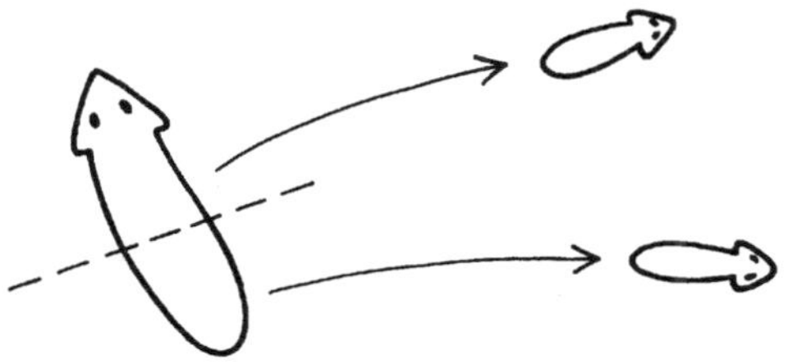

As Ted read about regeneration and cloning, he found that starfish also have great powers of regeneration. As a matter of fact, commercial fishermen in the North Atlantic would, at one time, chop the starfish into tiny pieces to kill them so that they would not clog up the fishing nets. The fishermen were very discouraged to see the starfish population increase rather than decrease after they were chopped into pieces—and you know the reason why! In his reading, Ted also found information about an award-winning high school science fair project that showed that vitamin A increased regeneration in starfish. After reading about this successful science fair project, Ted was motivated to **test the effect of vitamin A on regeneration in planaria** in his science fair project.

1. Based on all the information Ted has available to him, write a narrow question that Ted could answer through a science investigation project. (Be sure that this question meets all of the criteria listed on page four.)

____

2. Choose one of the things that you are interested in but cannot fully explain from page three; use it to write a narrow question that you could answer through a science investigation project. (Be sure that this question meets all of the criteria listed on page four.)

____

You now have at least one **clear and manageable question that is suitable for your project**. As other ideas come to mind, you should write them down in your **science investigation journal**. (The use of the journal will be explained later.) At some point, you will be able to select the one question that is **most interesting to you** for your science investigation project.

© Mark Twain Media, Inc., Publishers

Project Feasibility Study

Once you have settled on a good question to investigate in your science project, you will need to be certain that you have all of the necessary things to complete the project. This can be accomplished by doing a **project feasibility study** that covers three main aspects of science investigation projects: time, supplies, and information.

As for **time**, if you are going to treat and germinate seeds, you may only need several days before you are able to record your results. Finishing the experimental part of a planarian regeneration experiment would not take any longer. However, if your project requires that you germinate and grow bean seeds to the flowering stage, then you would need to allow several weeks. You must also allow some time (at least two to three weeks) to complete the final paper and/or display board.

The **supplies** needed to complete any science investigation project should be identified and obtained before the project begins. Investigating the topic question about "the effect of nicking the seed coats of honey locust seeds on germination" requires a modest amount of supplies, including: (1) honey locust seeds (recently produced to ensure that the seed coats have not been affected by weathering), (2) a file (to scar the seeds), and (3) a germination chamber (constructed of simple materials in a large jar or beaker). All of the materials needed for this investigation are easily obtainable and inexpensive. Therefore, a separate budget is not necessary.

Information that is available in books, magazine articles, and on the World Wide Web is very important to the development and interpretation of your project. It is possible that you may discover through your reading that the answer to your original topic question is well-known and not worth the time and effort it will take to use it for your science investigation project. The germination of compass plant seeds is known to require at least eight weeks of cold treatment (called **stratification**) and scarring of the seed coat. Testing the effect of cold treatment in this plant would result in nothing more than a demonstration of what is already known. On the other hand, there are some topics that are so unique that very little is known about them. Let's say you were interested in investigating the growth requirements of a rare species of orchid. The lack of information about this particular orchid could turn out to be a major obstacle to your project.

© Mark Twain Media, Inc., Publishers

Name ______________________________ Date ____________________

Student Exercise

Use the topic question that you described for question two on page eleven of the last section or any other clear and manageable topic question for the following activity.

1. Topic Question: (Write down your topic question.)

____

____

2. Project Feasibility Study

   a. Time (Describe the amount of time it will take you to do the various parts of your project.)

____

____

____

   b. Supplies (Describe the various supplies and equipment needed and where you can obtain these items.)

____

____

____

____

   c. Information (Describe the amount of information about your topic that you think will be available from the library and other sources. Circle one.)

   Abundant Adequate Little None

3. Is it feasible for you to do a science project on this topic? Why or why not?

____

____

The project feasibility study (PFS) should begin right after you have selected your topic question. However, it may not be possible to complete your PFS until you have gathered some information about your topic by searching through books, articles, and the World Wide Web.

© Mark Twain Media, Inc., Publishers

Name ______________________________ Date ______________

105 Ideas for Science Fair Projects

A list of 105 science fair titles that have been used by middle and high school students at science fairs and academies have been included in this section. Not every project on the list is in the form of a clear and manageable question, but you can usually derive the question (or questions) investigated by analyzing the project title. For example, the title "The Effect of Nicking Seed Coats on the Germination Rate of Honey Locust" implies the question "Does nicking the seed coats of honey locust affect their germination rate?". (This was an easy one!) Of course, there are some project titles like "Bacteria in Our Environment" that are very difficult to determine what questions have been researched in the science project. The above example could be talking about the different kinds and numbers of bacteria that are collected from environmental surfaces, or it could involve the effect of various disinfectants on bacteria. These cases encourage you to be creative and to come up with a longer list of possible questions for your own project.

Student Exercise

For each of the following student science project titles, write down at least one good question that might be answered through a science fair project.

1. "How Exercise Affects a Person's Blood Pressure"

____

____

2. "The Effect of Temperature on the Length of Time a Battery Lasts"

____

____

3. "A Comparison of Leaf Pigments in Different Plants"

____

____

4. "Spiders in Our Homes"

____

____

© Mark Twain Media, Inc., Publishers

Science Fair Project Title List

Here is a list of science fair projects for your consideration.

- Food preference in gerbils
- Does adding salt to water change the temperature at which it boils?
- The effect of temperature on how long a soap bubble lasts
- The effect of different light intensities on the growth of sunflower plants
- The effect of light on the growth of bread mold
- Which bird feed do birds like best?
- A comparison of calories in five different kinds of peanuts
- The effect of acid rain on the growth of Wisconsin fast plants
- The effect of acid rain on the germination of apple seeds
- Black walnut juice and marigolds to prevent weeds in your garden
- The effect of sugar water on the survival of cut flower stems
- The effect of salt on the growth of bean plants
- Hummingbird color preferences at feeders
- Bird feed consumption at different colored bird feeders
- Bacterial growth in apple juice and apple cider
- Does auxin affect seed germination?
- Can magnesium affect seed germination?
- Will acid rain affect the cell structure of spirogyra algae?
- The effects of surface area on fuel burn time
- Does the life of a light bulb depend on its wattage?
- Sudsy soaps
- The effect of wire coils on the strength of an electromagnet
- Packaging eggs and shock resistance
- The effect of fiction on velocity
- The effect of temperature on golf ball performance
- The effect of temperature on the rate of water absorption in cut carnations
- The antibiotic effects of bread mold on bacteria
- The effect of the color of light on the growth of sunflowers
- How changes in gravitropic responses affect plants
- A study of the abundance of prey available to arachnids through the use of artificial webs
- The feeding habits of winter birds
- The germination of gamma grass, *Tripsacum Dactyloides*

© Mark Twain Media, Inc., Publishers

- Hydroponics: growth for the future
- A comparison of the heat-conducting abilities of different metals
- Iron in your food
- The effect of a classroom seating arrangement on student performance
- Prejudices in children: When do they start?
- The golden ratio and its effect on heart rate
- The effects of a small magnetic field on the movement and behavior of laboratory mice
- The effect of paper airplane design on flight distance and flight time
- A comparison of the water content of different kinds of fruit
- The amount of fat in fast foods and store-bought hamburgers
- The effect of car exhaust fumes on the growth of plants
- An investigation of bacteria and fingernails
- The effects of the amount of water on the number of stomata in peas
- Acid rain: how it affects plant growth
- Using lichens to measure lead pollution along the Missouri River
- The effect of wing shape on lift
- An investigation of the mysteries of Fibonacci
- Does light affect population growth rate in euglena?
- The effects of radiation on pea seeds
- The effect of humidity on the behavior of isopods
- Are soap bubbles good for anything but a bath?
- Learning styles and memory retention
- The effects of car exhaust on seed germination
- Why do rocks sink and supertankers float?
- The effects of ultrasonic waves on the growth of peas
- A study of bridge construction
- Which brand of gum is the most viscous?
- The effect of design on the efficiency of a propeller
- The effects of heavy metals on the growth of Wisconsin fast plants
- The effects of ultraviolet light on the photosynthetic rate of soybeans
- The fermentation of yeast: optimal temperature and Ph
- The effects of water flow on a water wheel
- Day or night: When do amaryllis plants grow more?
- The effects of stress on the germination of corn seedlings
- The effect of rotation on fruit fly development
- Comparing the tensile strength of different metals
- A study of the ability of goldfish to learn a maze

© Mark Twain Media, Inc., Publishers

- The effects of microwave radiation on seed germination
- The effects of caffeine on the respiratory rate of cockroaches
- The effects of watching fish on blood pressure in people
- The effects of distractions on memory and learning
- The effects of music on the ability to memorize nonsense syllables
- The design and construction of a rigid sail model for all vessels
- An analysis of the nutrient content of breakfast cereals
- The use of pitfall traps to determine insect diversity
- The effects of fluorescent light on the learning abilities of white mice
- An analysis of the relationship between music and plant growth response
- An investigation into the effects of irradiation in the seed stage on the growth and development of marigolds
- The fruit of hedge apple: determination of a natural repellent of crickets
- Effect of electromagnetism on regeneration in planaria
- The effects of different colors of light on head regeneration in the brown planarian
- The effects of acid rain on regeneration in planaria
- Analysis of the effects of detergents on the heart rate and survival of daphnia
- The effects of nicotine on the heart rate of daphnia
- The effects of nicotine on the cell shape and survival of euglena
- The amount of vitamin C present in ordinary foods
- The effects of electric currents on germinating seeds
- Comparing the effects of antibacterial soaps on bacterial growth
- The antibacterial effect of common sauces
- Particulates in the air
- Is the purity of bottled water consistent with the claims that distributors make?
- A study of the effect of the plant hormone, auxin, on the growth of bean plants
- The effects of eye dominance on task performance
- The effect of age on successful mating in fruit flies
- Multiple intelligence in the career world
- The fungi around us
- The effects of acidity on metals
- Which form of insulation is most effective?
- How terraces help stop soil erosion
- Determination of vitamin C in aging fruit
- The effects of vitamin C on the visible characteristics of the wild type fruit fly
- Gender-based memory
- An investigation of the lung capacity of smokers and nonsmokers

© Mark Twain Media, Inc., Publishers

Name ______________________________ Date ________________

Student Exercise

1. From the list of science fair project titles, select five titles that would be considered physical science subjects and give the basic science category for each title. (These categories are discussed on page seven in case you have forgotten.)

Physical Science Project Titles	Basic Science Categories
a. ______________________	a. ______________________
______________________	______________________
b. ______________________	b. ______________________
______________________	______________________
c. ______________________	c. ______________________
______________________	______________________
d. ______________________	d. ______________________
______________________	______________________
e. ______________________	e. ______________________
______________________	______________________

2. From the same list, select five project titles that would be considered life science subjects and give the basic science category for each title.

Life Science Project Titles	Basic Science Categories
a. ______________________	a. ______________________
______________________	______________________
b. ______________________	b. ______________________
______________________	______________________
c. ______________________	c. ______________________
______________________	______________________
d. ______________________	d. ______________________
______________________	______________________
e. ______________________	e. ______________________
______________________	______________________

FINDING WHAT IS KNOWN ABOUT MY TOPIC

Searching the Library

Both **school** and **public libraries** are good places to start your search for information about your science investigation topic. As long as you are courteous and patient, the **librarian** can usually help you to find materials related to your topic. The librarian may have information of interest to you in a **vertical file**. The vertical file may be a folder of newspaper and magazine articles on a particular subject that was assembled by the librarian. This is why it is always wise to check with the librarian about your science investigation topic as soon as possible.

The next step would be to go to the **card catalog** where you search by **author, title,** and **subject**. The cards in the card catalog are usually arranged alphabetically in file drawers by author, title, and subject. Many libraries have made it easier for you to search for information about a topic by having the entire card catalog on a computer database. The librarian can show you how to use the card catalog or the computer database program.

Indexes such as the *Reader's Guide to Periodical Literature* and the *Children's Periodic Guide* use **subjects** to find current articles about a topic. **Encyclopedias** and **almanacs** are other good sources of information covering your science investigation topic.

After you have exhausted the sources about your topic at the local school and public libraries, you might want to examine some of the more advanced abstract journals available at a **college** or **university library**. **Abstract journals** contain summaries (called **abstracts**) of the **primary journal articles** where all original scientific research is published. Some of the major abstracting journals that can be used in student science investigations are *Bioabstracts*, *Chemabstracts*, and *Psychology Abstracts.*

The primary journal articles are **reviewed** and approved by experts in each scientific field before they are accepted for publication. Obviously, there must be a great number of primary journals to publish the tremendous number of scientific studies that are conducted in the life and physical sciences around the world each year. As of 1991, there were over 100,000 primary scientific journals being published. This represents significant progress since the first primary scientific journal, called *Journal des Scavans*, was published by the French Academy of Science in 1665!

The college or university library that you use may not have all of the primary scientific journals that you need, but for a small charge the **reference librarian** can often order your articles through interlibrary loan—as long as you can wait several days. Waiting for information might not be a problem if you **start your project early and budget your time properly**.

The use of **3″ x 5″ note cards** can be very helpful for organizing information from articles, books, and other sources. The following information should be obtained from any source about your science investigation topic:

1. title of the book or article
2. author
3. pages used (web page address for Internet information)
4. publisher of book or name of magazine/journal
5. date of publication
6. summary and other notes about the book or article

The use of separate 3″ x 5″ note cards to record information from each reference source is recommended, because these cards can later be organized in various ways to help you in writing up your paper or display. Below are two examples of reference cards based on a book and a magazine article.

Book (Front):

1. The Invisible World
2. by John Walker
3. Pages 44 to 64
4. Riley and Sons Pub. Co.
5. 1989

Book (Back):

6. This book was about the many different kinds of microorganisms. Chapter six discusses the life cycles of various pond invertebrates, including "water bears."

Article (Front):

1. The Effect of Light Intensity On Movement In Planaria
2. by Janet Reed and Joseph Meeks
3. Pages 31 to 37, Vol. 6, No. 3
4. Flatworm Transactions
5. 1993

Article (Back):

6. The experiments in this paper show that planaria avoid strong light and move into dark areas.

© Mark Twain Media, Inc., Publishers

Searching the World Wide Web

The **World Wide Web** can be another great source of information about your science investigation topic. In order to "surf" the World Wide Web, you need a computer that is linked to the Internet and a **web browser program**. Once you are on the World Wide Web, you can select a **search engine**, such as *Yahoo*, *Altavista*, or *Web Crawler*, and type in your topic ("planaria," in our example), as shown below.

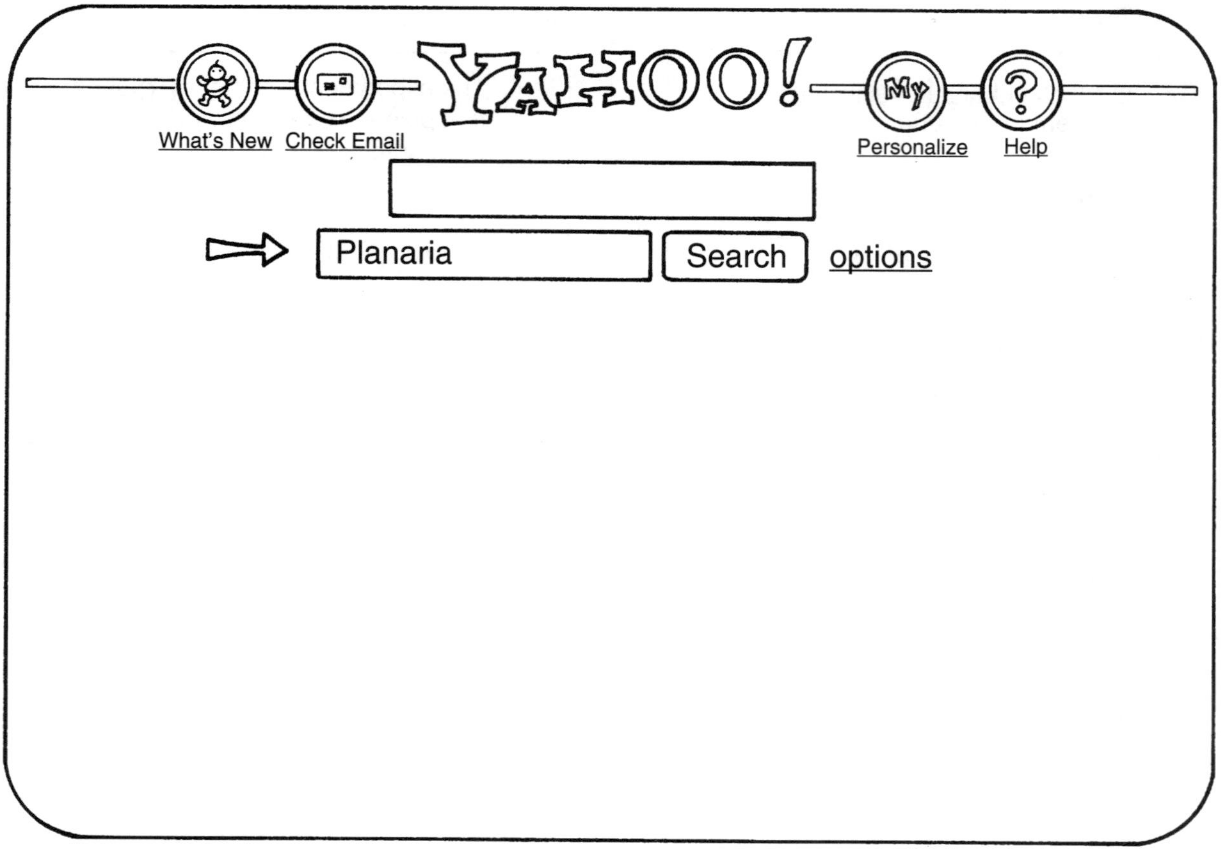

The amount of information available on the World Wide Web is huge, and it is expanding each day.

You should also fill out reference cards with all six standard pieces of information for each **web site** that you used. Be sure to record the **web site address** on the reference card, along with the pages in the article as information item #3. Below is an example of a web site on mealworms with the arrow pointing to the web site address. (You can also print out whatever is on the monitor screen.)

Search Result Found **815** web pages for **mealworms**

Yahoo! Pager

Categories	Web Sites	**Web Pages**	News Stories	Net Events
Other suggested spellings: meal worms				

Web Pages (1-20 or 815)

- cricket Home Page - MILLBROOK CRICKET FARM 550 Old Highway 49 South Richland, MS 39218 Phone: (601) 932-9333 Toll Free: (800) 654-3506 Fax: (601) 939-5315 EMAIL AT: ..milbrook@teclink.net Cricket's .. Pet Food .. Guaranteed Live Delivery .. Buy Direct We accept Visa, .
--http://www.jwaltman.com/

Web Site Address

- Rainbow Mealworms, Inc. - Rainbow Mealworms, Inc. 126 E. Spruce St. Compton, CA 90220 Orders: (800) 777-9676 Inquiries: (310) 635-1494 Please mention this WWW page when you call "Birds Love Mealworms" [Mealworm Pricing] Bulk Sizes: Mini, Small, Medium, Large & Mixed 2 ...
--http://www.herp.com/rainbow/

If you cannot find the author or exact source for web site information, you should not use the information as a reference in your final report or display.

If you really get interested in using the World Wide Web, you can go to various **on-line forums** and **web boards** that permit you to communicate about your topic with other interested computer users. You should ask your teacher, school librarian, or any other knowledgeable computer user to show you how to **"surf the web."**

© Mark Twain Media, Inc., Publishers

Name ______________________________ Date ____________________

Student Exercise

1. Go to your school (or public) library and use the card catalog (or computer database program, if available) to find two books that are related to the topic question you used to answer question two on page eleven. (You may use any other clear and manageable question if you wish.) Fill out the first five pieces of information (1. Title, 2. Author, 3. Pages of the book used, 4. Publisher, and 5. Date of Publication) in the sample reference cards below.

Reference Book 1

1.
2.
3.
4.
5.

Reference Book 2

1.
2.
3.
4.
5.

2. Now use either the *Guide to Periodical Literature* or the *Children's Periodical Guide* to find two current magazine articles that are related to the topic question that you used in the above question. Fill out the first five pieces of information about these articles in the sample reference cards below. (Use the total number of pages in the article for 3. Pages.)

Reference Article 1

1.
2.
3.
4.
5.

Reference Article 2

1.
2.
3.
4.
5.

© Mark Twain Media, Inc., Publishers

Name ________________________ Date ______________

3. Use a computer that has an Internet browser program to find two World Wide Web sites that are related to your topic. Fill out the first pieces of information in the sample reference cards below. Be certain to record the web site address in information piece number three.

Web Site Address 1

1.
2.
3.
4.
5.

Web Site Address 2

1.
2.
3.
4.
5.

4. What is the name of the librarian and the phone number of the library where you completed this activity?

____

____

© Mark Twain Media, Inc., Publishers

Contacting Experts

In exploring the libraries and the World Wide Web, you may find a book, article, or web site source that is especially close to your topic. Writing the author of this work can result in more information and insights into your science investigation question.

Before writing any expert, you should be certain that you have done a thorough library search and are easily able to talk about your science project. You should not expect a professional scientist to tell you what to do in your project, but rather to give you ideas and methods about the research that he or she is doing. Professional research scientists are often very busy people, but they are usually glad to help a student who is sincerely interested in their area of research. Below is an example letter from a seventh grade student to a research scientist.

To: Dr. Elmer T. Baumann
Department of Biology
University of Wisconsin
Madison, Wisconsin

Dear Dr. Baumann:

My name is Ted Jones, and I am a seventh grade student at Truman Middle School in Middlebury, Iowa. During the past several weeks, I have been working on a science investigation project dealing with the effect of vitamin A on regeneration in planaria. I have read your paper titled "Vitamin Enhancement in Animal Tissue Cultures" that was published last year in The Tissue Culture Journal. *I found your paper to be very interesting and closely related to my project. Can you suggest any further sources of information that might be helpful to me?*

I know that you are extremely busy with your college teaching and research, but any of your suggestions will be greatly appreciated. Please send your response to Ted Jones, 2413 Maple Lane, Middlebury, Iowa.

Thank you for your consideration.

Sincerely yours,

Ted Jones

Don't be afraid to put additional information about yourself and your science investigation project into the letter.

If the research scientist whom you are writing is from your local area, a telephone call might be more effective than a letter. Of course, the telephone conversation should at least cover the same information that would be in the letter.

Contacting a research scientist is not necessary for the development of a good science investigation project, but it can be helpful in some cases.

© Mark Twain Media, Inc., Publishers

MAKING A HYPOTHESIS AND DESIGNING A STUDY

Turning My Question Into a Hypothesis

Five examples of topic questions for student science fair projects are listed below.

1. What effect does water temperature have on the size of soap bubbles?
2. What effect does hot water have on the germination of geranium seeds?
3. What effect does the color of light have on the growth of radish seedlings?
4. Which type of seeds do house finches like best?
5. Does the life of a light bulb depend on wattage?

Even though many forms of the scientific method are used, all science investigations are based on hypotheses that are tested in order to answer questions about the material world. A **hypothesis** is essentially a topic question that has been reworded into a form that can be tested. In the example below, question one has been restated in the form of a hypothesis.

A. Hypothesis: Warmer water will produce larger bubbles than cooler water.

© Mark Twain Media, Inc., Publishers

Name ______________________________ Date ______________________

Student Exercise

1. Change topic questions two, three, four, and five from the examples on page 27 into hypotheses.

   B. Hypothesis: ______________________________

   ____

   ____

   C. Hypothesis: ______________________________

   ____

   ____

   D. Hypothesis: ______________________________

   ____

   ____

   E. Hypothesis: ______________________________

   ____

   ____

It is possible to have more than one hypothesis for any topic question. Two **alternative hypotheses** for Hypothesis A are: (1) Cooler water will produce larger soap bubbles than warmer water, and (2) Water temperature has no effect on the size of soap bubbles. However, in designing your experiment, just use the **one hypothesis** that is most likely to be supported.

© Mark Twain Media, Inc., Publishers

Types of Research Studies (Descriptive, Correlation, and Experimental)

The following three basic types of research studies are often recognized: descriptive, correlation, and experimental. All of these scientific studies are very good ways to find out new information about the material world, but experimental studies are generally considered the most productive in modern science.

Descriptive studies describe the way things are in the material world without trying to explain their causes. If a student studies the variation in size (length and width) of cottonwood tree leaves, then she or he is doing a descriptive study. In another descriptive study, a student might study the density of tiny breathing pores (called **stomata**) on the surface of the leaf. In neither of the above studies is there an attempt to establish the cause (or causes) of leaf size or stomatal density.

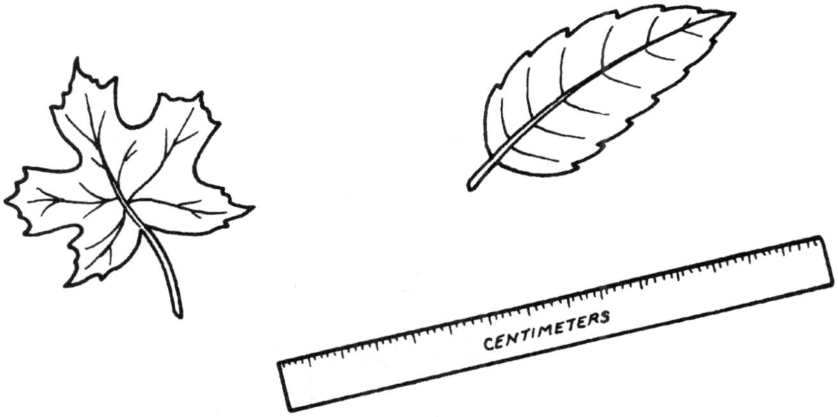

Let's suppose that another student studied both leaf size (length and width in centimeters) and stomatal density (stomata/cm^2) to see if there is any connection between the two. This student might eventually find that smaller leaves tend to have greater densities than larger leaves. This type of research is called a **correlation study** because it makes a connection between two separate observations but does not establish any causes. When a **herpetologist** (reptile and amphibian expert) studies and compares the snakes found in a forest and a nearby prairie, she or he is also doing a correlation study.

Aristotle

Both descriptive and correlation studies are very important scientific research approaches that have been used for hundreds of years, especially in geology and in the life sciences. **Aristotle**, the ancient Greek natural philosopher who is considered the "father of biology," used descriptive and correlation studies over 2,000 years ago to discover the life cycles of many interesting marine animals, such as octopuses and cuttlefish. Many new species of plants and animals are currently being discovered in the tropical rain forest through descriptive and correlation studies.

© Mark Twain Media, Inc., Publishers

Experimental studies consider all factors (sometimes called **variables**) that might change an experiment in order to establish the cause of observations. Suppose that a student (let's call him Jim) has selected the topic question, "Does wind have any effect on the water loss from the leaf surface of Zebrina plants?". From this question, Jim formulated the following hypothesis: "Wind on the leaf surface of Zebrina plants will increase the rate of water loss." (This water loss from the leaf is called **transpiration**.) Jim's hypothesis was based on the physical principle that wind tends to increase the evaporation of water from a solid surface.

Testing My Hypothesis

According to Jim's hypothesis, the wind causes increased transpiration, but there are also other variables such as light, temperature, and humidity that might affect transpiration. If Jim is going to perform a successful science experiment, he must **identify all of the important variables in the experiment and control them** so that he can test the effect of just wind on transpiration. Jim must also have an accurate way to **measure** transpiration.

In Jim's experiment, wind is considered the **independent variable** because Jim is free to set the wind speed and time that the plants will be subjected to the wind. (Jim will use a fan to produce the wind **treatment**.) The **dependent variable** is what is measured—the amount of water lost from the Zebrina plant, in this example. Transpiration from the leaf can be easily measured with a device called a **potometer**.

The potometer setup for Jim's experiment is shown below.

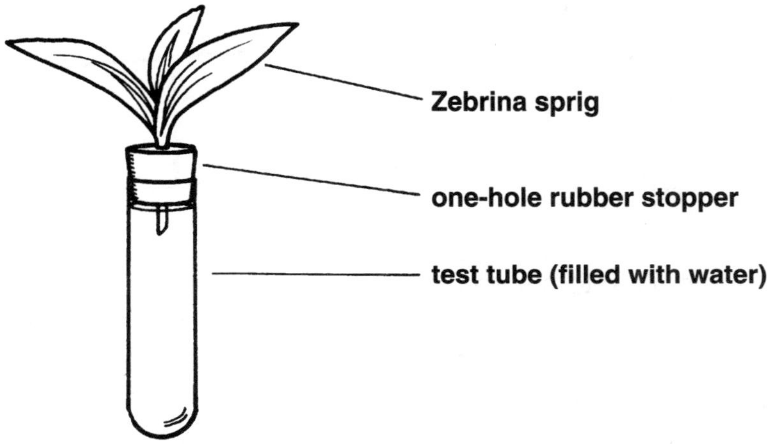

As transpiration takes place, water evaporates from the leaf surface through tiny pores (stomata), and the Zebrina plant sprig takes in the same amount of water from the test tube as was lost through transpiration. The level of water in the test tube goes down and shows how much water was lost by transpiration (shown below).

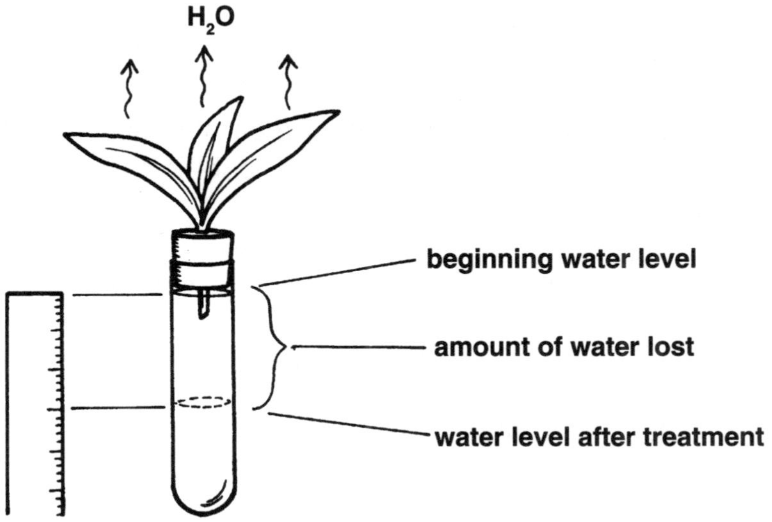

The dependent variable (water loss from the plant) can be easily determined by measuring the distance that the water level drops in the test tube during the experiment. (Jim could also figure out a way to determine the water loss by weighing.)

In Jim's experiment, he decided to treat 10 separate plants in a test tube rack with wind produced by the medium speed on a small fan for one hour. This was his **treatment group** (shown below).

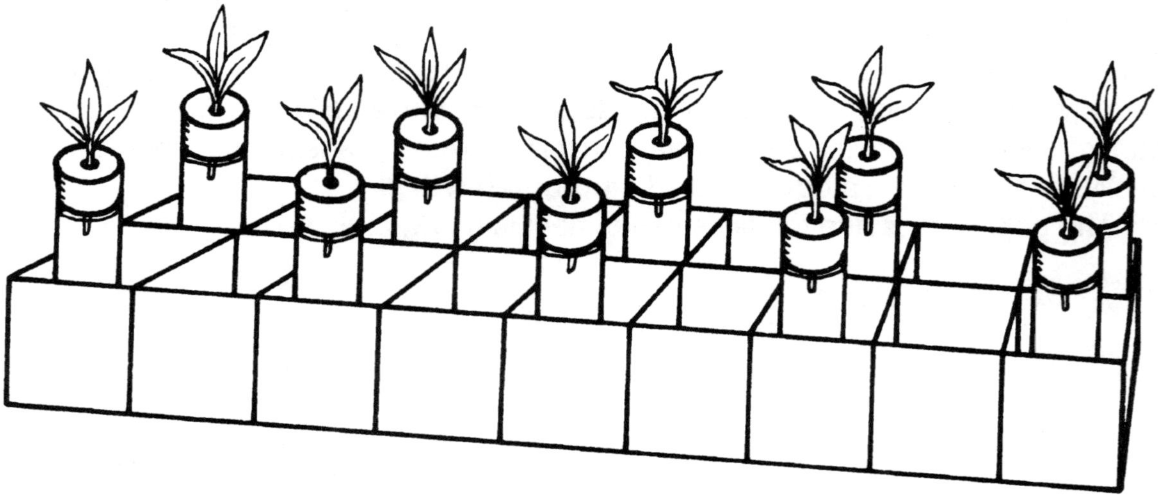

In addition to his treatment group, Jim set up a **control group** of 10 Zebrina plants. Jim saw to it that the plants in the control group were subjected to the **same conditions** of light, temperature, and humidity as the plants in the treatment group. The only difference was that the treatment group plants received wind treatment and the control group plants did not. This way, Jim could assume that any difference in transpiration between the two groups was **caused** by the wind—the dependent variable.

Earlier in this section, the following hypothesis was discussed:

A. Hypothesis: Warmer water produces larger bubbles than cooler water.

In designing an experiment based on this hypothesis, water temperature (warm or cool) will be the **independent variable**. Other variables, such as the concentration of the soap, how the soap is mixed, light, and air humidity, might affect bubble production and should be controlled. The size of the bubbles is the **dependent variable**—what is measured.

Name ______________________________ Date ________________

Student Exercise

1. Determine the independent and dependent variables for Hypotheses B, C, D, and E from your answer to question one (page 28). List the important variables to be controlled for each hypothesis.

   B. Hypothesis: ______________________________

   ____

   Independent Variable: ______________________________

   Dependent Variable: ______________________________

   Factors to be Controlled: ______________________________

   ____

   ____

   C. Hypothesis: ______________________________

   ____

   Independent Variable: ______________________________

   Dependent Variable: ______________________________

   Factors to be Controlled: ______________________________

   ____

   ____

   D. Hypothesis: ______________________________

   ____

   Independent Variable: ______________________________

   Dependent Variable: ______________________________

© Mark Twain Media, Inc., Publishers

Name ______________________________ Date ______________

Factors to be Controlled: ______________________________

____

____

E. Hypothesis: ______________________________

____

Independent Variable: ______________________________

Dependent Variable: ______________________________

Factors to be Controlled: ______________________________

____

____

2. Select a topic question and hypothesis that you are interested in using for your science fair project. Provide the following information about your topic.

Topic Question: ______________________________

____

____

Hypothesis: ______________________________

____

____

Independent Variable: ______________________________

Dependent Variable: ______________________________

Factors to be Controlled: ______________________________

____

____

© Mark Twain Media, Inc., Publishers

area, length by width, of all leaves on each plant and assign sprigs to control and treatment groups so that each group has the same overall leaf size.)

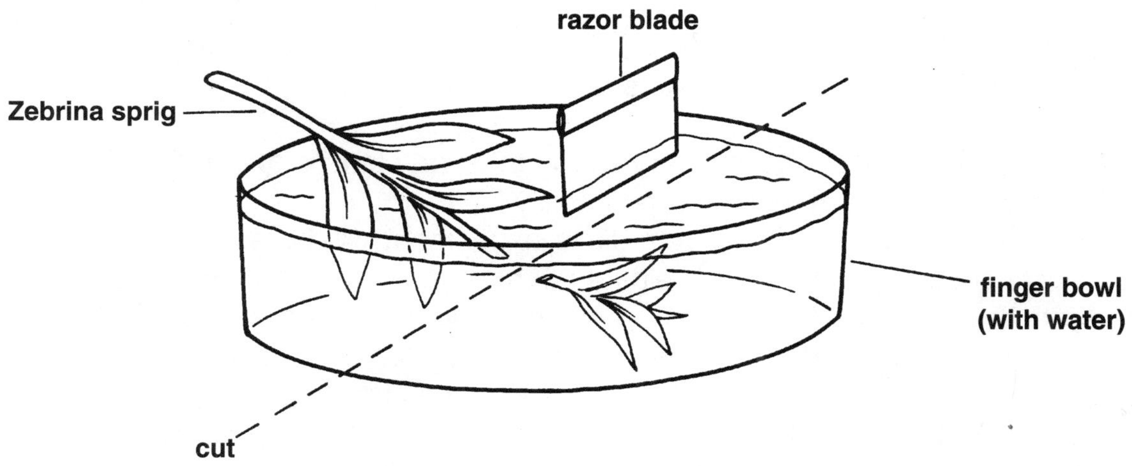

Step 4: Fit the one-hole rubber stopper into the test tube so that it fits tightly. (Some water will spill out as this is done.)

Step 5: Add petroleum jelly around the stem and rubber stopper as shown below to seal any air leaks.

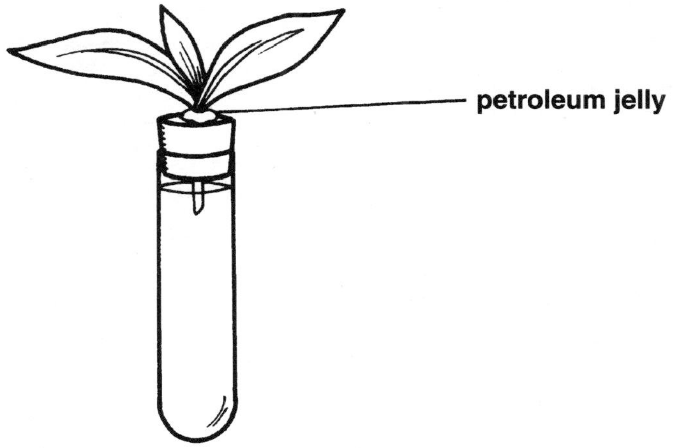

Step 6: Repeat steps three through five for all 20 test tubes to be used in this experiment.

Step 7: Place both of the test tube racks filled with plant cuttings under the same conditions of light, temperature, and air humidity. (Measurements of these variables could be made if the necessary measuring equipment is obtained.)

Step 8: Place a fan near one of the test tube racks of plants and turn it on medium speed for one hour. These plants will be the **treatment group**. The other test tube rack of plants (the **control group**) will be placed somewhere behind the fan and will have all variables exactly the same as the treatment group, except that this group will not receive the wind treatment from the fan.

© Mark Twain Media, Inc., Publishers

Step 9: After one hour, the water levels of all test tubes should be measured with a metric ruler, and this **data** should be recorded in the student journal.

Step 10: The entire experiment should be repeated at least two times. (Switch treatment and control group plants for each trial.)

Repeatability of results is a very important characteristic of any scientific experiment. The trials help to demonstrate repeatability.

It is assumed that the data from Jim's experiment will be summarized and that conclusions about his hypothesis will be made within a reasonable time frame. If Jim thinks that he might have difficulty completing everything on time, then he should add a final step (11) that tells exactly when the final report and display boards are due.

The project proposal described above should be Jim's first entry into his student project journal. A **student project journal** is a separate, permanently-bound notebook that is reserved for notes and data from the science investigation project.

Student Exercise

In Question two of this section (page 34), you selected a topic question and formulated a hypothesis about a subject that you are interested in investigating. Use this topic (or another one if you have changed your mind) to **write up a project proposal in your own student project journal.**

Have your teacher check this proposal before you begin working on the project. At this point, you have come a long way in your science investigation project. **Congratulations!**

MAKING SENSE OUT OF MY PROJECT RESULTS

As you carry out the steps of your science investigation, all of your observations and any other related ideas that come to mind should be written into your **student project journal**. There are many good ways to enter information into the project journal, but it is always wise to put the date of your entry at the top of each page. This will help you to organize your results when it is time to write your final paper.

Tables and Averaging

The most important information in your project journal is the **data** that you obtain from the observations and measurements made during the investigation. For example, in Jim's experiment, which was explained in the last section (pages 36-38), he compared the amount of water loss (transpiration) between a group of 10 Zebrina plants that were treated with wind and 10 Zebrina plants that were not treated with wind. As Jim measured the water loss from each plant, he needed a way to **organize** his **raw data** in his project journal. Tables are generally used for this purpose. Jim's results for two separate **trials** are given in the **raw data table** below.

Plant	**Trial No. 1** Treatment Group (mm)	**Trial No. 2** Control Group (mm)
#1	16	1
#2	12	1
#3	13	1
#4	14	0
#5	15	2
#6	17	1
#7	11	5
#8	13	4
#9	14	0
#10	14	2
Average	13.9	1.7

© Mark Twain Media, Inc., Publishers

Plant	**Trial No. 1** Control Group (mm)	**Trial No. 2** Treatment Group (mm)
#11	1	14
#12	2	15
#13	2	16
#14	2	13
#15	2	10
#16	4	17
#17	1	11
#18	0	12
#19	4	11
#20	1	13
Average	1.8	13.2

Notice that Jim has provided a place for the averages of each treatment (wind versus no wind). **Averaging is a way to get a single quantity that represents a group of several quantities.** In Trial #1 for Jim's treatment group, water losses of 16, 12, 13, 14, 15, 17, 11, 13, 14, and 14 millimeters are represented by the average quantity of **13.9 mm**. The water losses from the control group in Trial #1 are 1, 2, 2, 2, 2, 4, 1, 0, 4, and 1, for an average of **1.8 mm**. In Trial #2, Jim's averages were **13.2 mm** for the treatment group and **1.7 mm** for the control group.

The data for both of Jim's trials are summarized in the **following table of averages**.

Average Water Loss (mm)

	Treatment	Control
Trial No. 1	13.9	1.8
Trail No. 2	13.2	1.7

The above summary table clearly shows a greater water loss (transpiration) with wind (13.9 mm and 13.2 mm) over no wind (1.8 mm and 1.7 mm). Special statistical tests can also be used to compare averages, but such tests are beyond the scope of this book. If you decide to use statistical tests, be certain that you fully understand how and why they are used. Your science teacher and any science fair or science academy judge will ask you to explain them.

© Mark Twain Media, Inc., Publishers

Bar Graphs

Another way to analyze and compare totals such as those from Jim's experiment is with a bar graph. A **bar graph** is usually used to compare total amounts.

Water Loss in Zebrina
Wind Treatment Versus No Wind

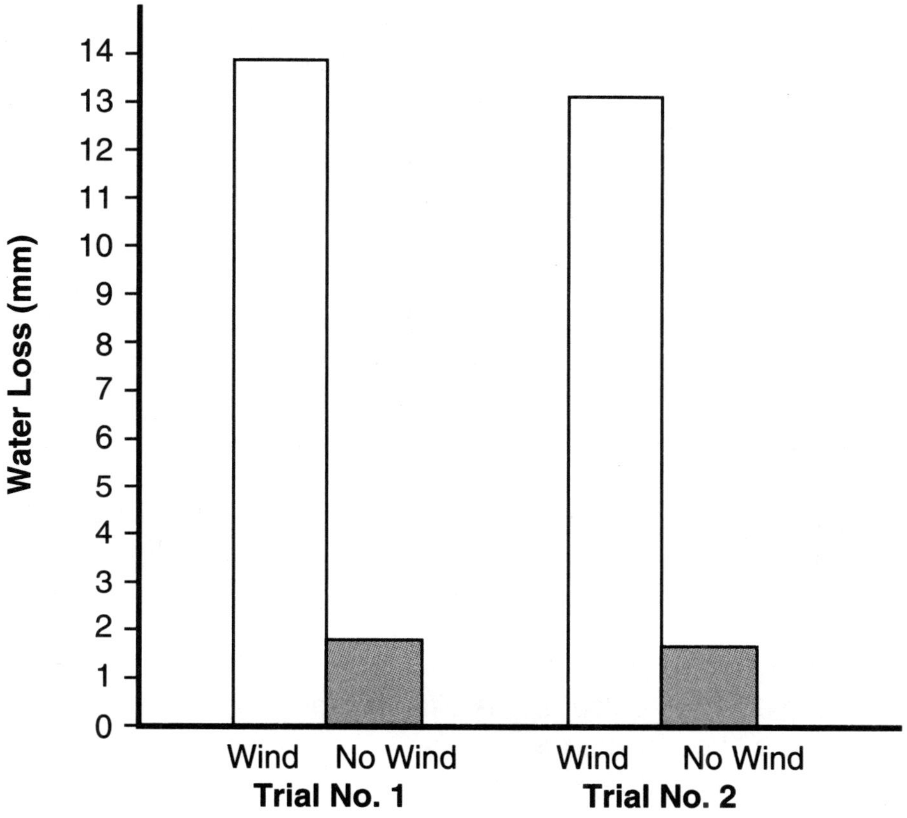

Name ______________________________ Date ________________

Student Exercise

Let's try a class activity as a way for you to review some of the things you have recently learned about recording and summarizing data. This experiment is based on pulse rate as the dependent variable. A person's pulse rate is fairly easy to determine. Put two fingers on your partner's wrist, as shown below.

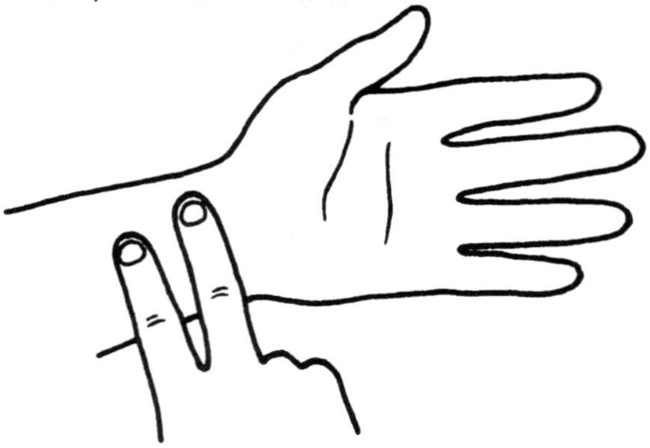

Press lightly with your two fingers to feel your partner's "pulse." (Your partner should be seated and relaxed.) The pulse is the result of increased blood movement in the radial artery of the wrist and is caused by a heartbeat. Therefore, the number of pulses is equal to the number of heartbeats.

1. In the space below, record the number of pulses per minute of your lab partner. This can be done by counting the number of pulses in 15 seconds and multiplying this number by four.

____

2. Do you think that the average pulse rate is different for people sitting and standing? State your answer as a hypothesis.

   Hypothesis: ____

____

3. Determine your lab partner's pulse rate while standing and record it in the space below.

____

4. Use the following raw data table to record the results for all of your classmates. Calculate the raw average pulse rates for sitting and standing. (This table could be put on the chalkboard and used for class discussion.)

Name ______________________ Date ____________

Student	Pulse Rate (pulses/minute) Sitting	Standing
1.		
2.		
3.		
4.		
5.		
6.		
7.		
8.		
9.		
10.		
11.		
12.		
13.		
14.		
15.		
15.		
16.		
17.		
18.		
19.		
20.		
21.		
22.		
23.		
24.		
25.		
AVERAGE:		

Name ______________________________ Date ____________

5. Do the results of your experiment support or reject your hypothesis?

____

6. Was the above experiment well-controlled? Why or why not?

____

____

____

7. Sketch a bar graph that compares the average sitting and standing pulse rates of your classmates in this experiment.

8. Design an investigation to determine if there is a difference in the average pulse rates between girls and boys in your class. In the space below, write down your initial hypothesis and a bar graph representing a summary of your class results. (Raw data from this investigation could be put on the chalkboard and used for class discussion.)

Hypothesis: ______________________________

____

Bar Graph:

9. Did the results from this investigation support or reject your hypothesis? Explain your answer.

____

____

____

© Mark Twain Media, Inc., Publishers

Line Graphs

Let's assume that a class did a science investigation to determine if longer hands go with longer feet. The table below contains the raw data from this investigation after all measurements were made.

Student	Hand Length (cm)	Foot Length (cm)
Mary	14	16
Amy	15	17
John	16	19
Kim	10	12
Tom	17	21
Bill	13	14
Don	10	13
Teresa	12	14
Judy	13	15
Dave	15	18
AVERAGE	13.66	15.92

The averages of hand and foot size from this table could be calculated and represented by a table or a bar graph, but this would not answer the question of whether longer hands go with longer feet. A better way to represent this data and answer the question would be with a line graph. A line graph shows how two factors change in relation to one another. The data from this investigation is summarized in the line graph below. Each point on the graph represents the hand and foot length of one student in the class.

Comparing Students' Foot Length to Hand Length

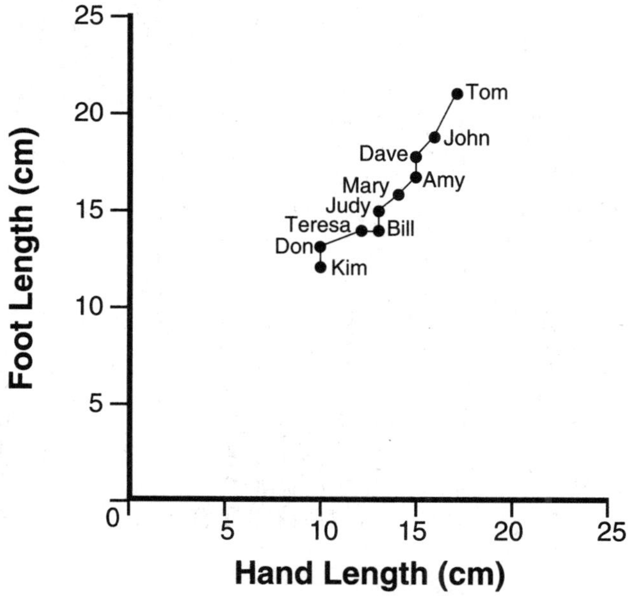

© Mark Twain Media, Inc., Publishers

In the line graph, the dots are connected to show a trend of increasing foot length with hand length. (This could also be stated the other way around.) As shown below, a straight line through the middle of all of dots averages the variation and clearly shows a steady increase of foot length with hand length.

Comparing Students' Foot Length to Hand Length

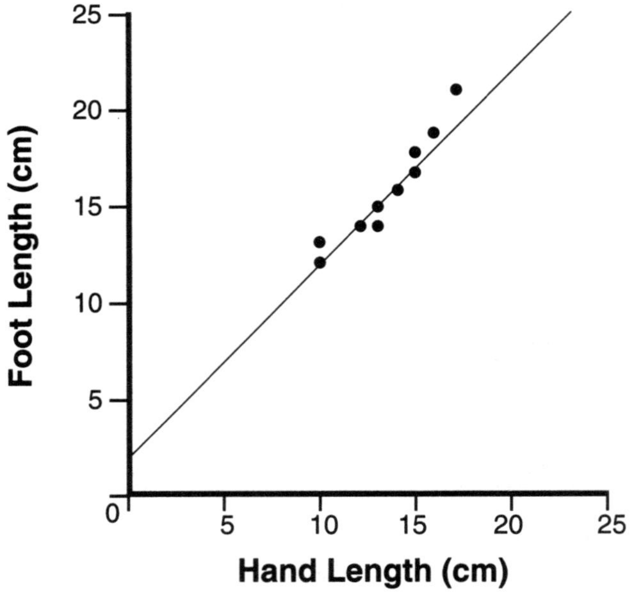

We can say that hand and foot length are positively correlated. Since neither of the above graphs about hand and foot length suggest that hand length and foot length cause one another, this investigation would most correctly be considered a correlation study and not an experiment. Please remember that correlation studies are valid science investigations, but they are not as highly regarded in science as experiments because they do not attempt to establish causes for observation.

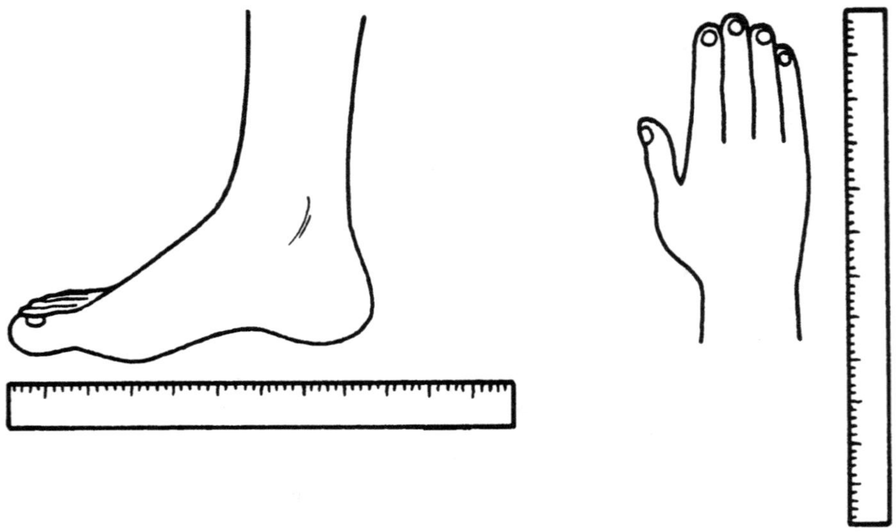

Name ______________________________ Date ________________

Student Exercise

1. Pick two body parts other than the two used in the example on page 46 and design a correlation study to see if the sizes of your two factors are positively correlated. Obtain your measurements from 10 other students in your class.

   a. What is your hypothesis? ______________________________

   ____

   b. Table of raw data:

   c. Line graph:

   d. Does your data support or reject your hypothesis? Explain your answer.

   ____

   ____

   ____

© Mark Twain Media, Inc., Publishers

Changes Over Time

Most line graphs consider that some factors may change with time. Let's consider another problem. Suppose a student named Jane asks the question: "Is there a difference in the way that shoe sizes of middle grade girls and boys change through the fifth, sixth, seventh, and eighth grades?". In order to answer this question, Jane obtains the shoe size of all fifth, sixth, seventh, and eighth graders in her school. She then averages the shoe size for girls and boys separately in each grade. Jane's summary table that shows the average shoe size for girls and boys in each grade is given below.

	Shoe Size	
Grade	Girls	Boys
Fifth	3.9	4.0
Sixth	4.2	4.4
Seventh	5.1	6.2
Eighth	6.0	7.0

A better way to show the comparisons of average shoe sizes between girls and boys among the fifth, sixth, seventh, and eighth graders would be with the following line graphs.

Girls' and Boys' Shoe Sizes in Fifth Through Eighth Grade

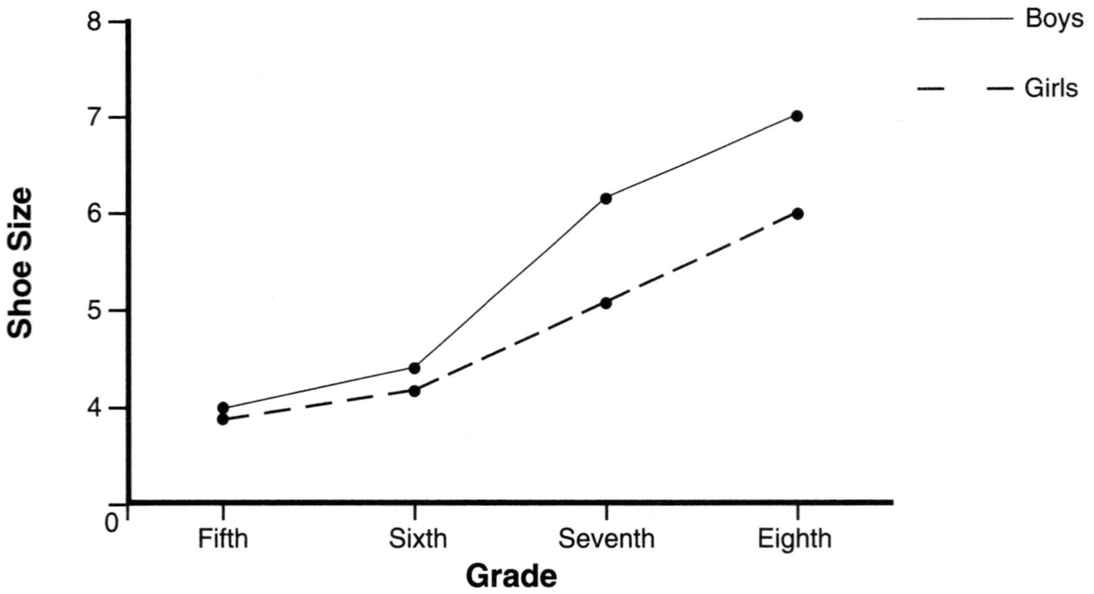

The girls and boys are represented by separate lines. Jane's line graphs clearly show a difference between the shoe sizes of girls and boys in each grade. Jane's line graphs also show a difference in how shoe size changes from the fifth through the eighth grades.

© Mark Twain Media, Inc., Publishers

Name ______________________________ Date ____________________

Student Exercise

1. Record the outdoor temperature in degrees Fahrenheit every hour throughout the entire school day. (This could be recorded on the chalkboard and done as a class project.)

a. Record your temperature measurements in the table below.

Time	Temperature (F)
9:00 A.M.	
10:00 A.M.	
11:00 A.M.	
12:00 P.M.	
1:00 P.M.	
2:00 P.M.	
3:00 P.M.	

b. Develop a line graph based on the data in the table above.

c. Does your data produce a straight line on the graph? (Circle one.)

Yes No

d. At what time was the lowest temperature recorded?

____

e. At what time was the highest temperature recorded?

____

Pie Charts

It is important to realize that not all line graphs will show a straight line. If you were to record outdoor temperatures every hour from dawn until dusk, your readings would usually go up throughout the morning and early afternoon and then come back down again toward dusk. You might want to test this "hypothesis."

Another way that is often used to show results of a study as a single picture is a pie chart. A **pie chart** shows the percentages of all possible observations. For example, suppose a student in a class of 20 students gets the following results:

Hair Color	Number of Students	Percentage
Black	8	40
Brown	6	30
Blond	4	20
Red	2	10

To find the percent, divide the number of the item you are focusing on by the total number of items and then multiply by 100. For example, $8 \div 20 = 0.4$, $0.4 \times 100 = 40\%$.

Since a circle is composed of 360 degrees and each slice of the pie chart is a portion of that 360 degrees, we must compute the number of degrees each slice will take from the total of 360 degrees. To do this, multiply the decimal form of the percentage by 360. For example, the decimal form of 40% is 0.40 ($40 \div 100$), so $0.40 \times 360 = 144$. There will be 144 degrees in the slice that represents black hair.

To actually draw the pie chart, draw a circle with a compass and find the center of the circle. Then draw a radius from the center to the edge of the circle. Place a protractor at the center of the circle and along the radius. Next make a mark where 144 degrees is on the edge of the circle. Draw a radius from the center of the circle to this mark. That completes the first slice. To draw the next slice, place your protractor on the last radius drawn and complete in the same way. Continue until you have drawn all the slices. You may want to shade each slice in a different way so they can be easily distinguished.

The pie chart for the above results would be:

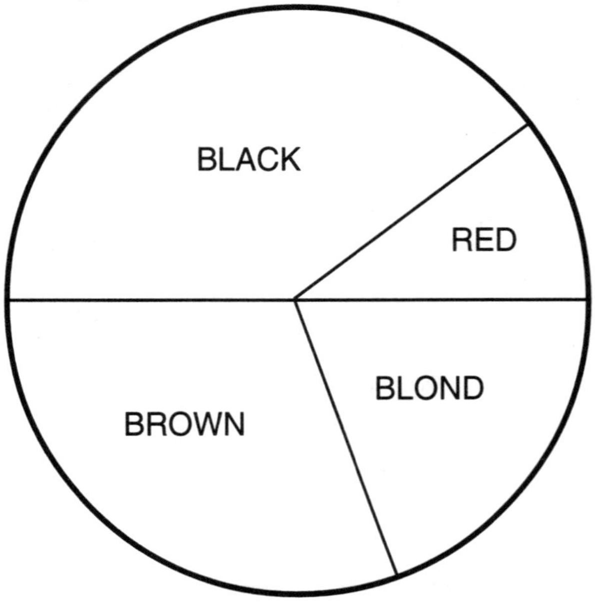

Name ______________________________ Date ______________

Student Exercise

1. In the space below, record your survey results and develop a pie chart to show the percentages of three different eye colors (brown, green, and blue) of students in your class. In cases where it is difficult to determine between two colors, select the one that is the closest fit.

   Raw Data Table:

Student	Eye Color	Student	Eye Color
1.		11.	
2.		12.	
3.		13.	
4.		14.	
5.		15.	
6.		16.	
7.		17.	
8.		18.	
9.		19.	
10.		20.	

Percentage of All Students:

Brown__________ Blue __________ Green __________

Pie Chart (draw sections and label):

Name ______________________________ Date ____________________

2. Record your survey results and develop a pie chart that shows the percentages of students with earlobes and attached ears. (This trait is illustrated below.)

Raw Data Table:

Student	Earlobes	Attached	Student	Earlobes	Attached
1.			11.		
2.			12.		
3.			13.		
4.			14.		
5.			15.		
6.			16.		
7.			17.		
8.			18.		
9.			19.		
10.			20.		

Percentage of All Students: Earlobes __________ Attached __________

Pie Chart (draw sections and label):

Expect the Unexpected

The importance of making careful observations and "expecting the unexpected" during your science investigation cannot be stressed enough! Even observations that seem unrelated to your topic question may have very important consequences. This was demonstrated in 1929 by Sir Alexander Fleming's discovery of penicillin. While culturing a common bacterium called Staphylococcus, Fleming noticed that some of his culture plates were contaminated by Penicillium mold. Many investigators would have quickly discarded the contaminated plates, but Fleming took the time to observe that the Penicillium mold was preventing the Staphylococcus from growing near it.

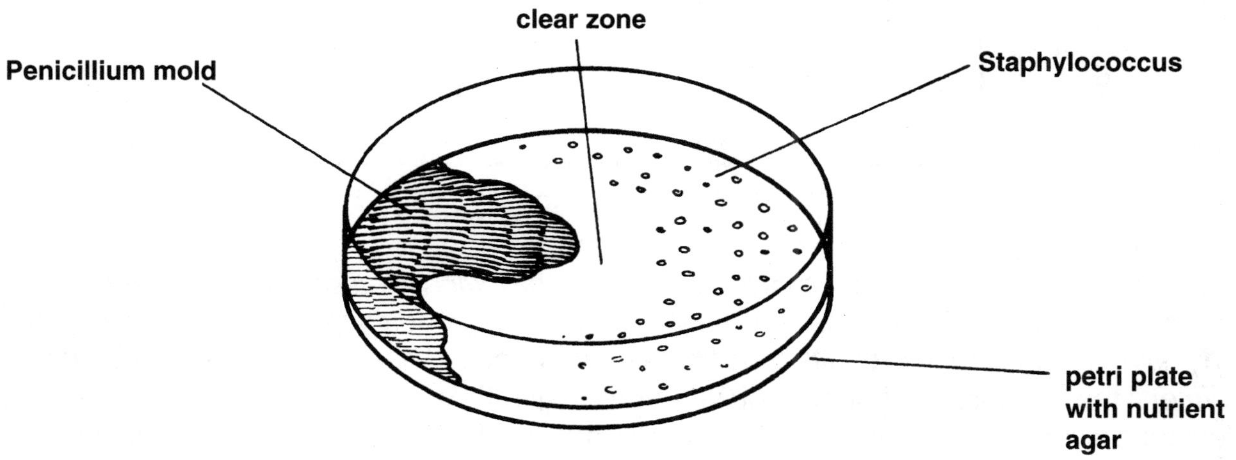

Fleming reasoned that if the bacterial killing agent from the mold could be isolated that it might be used to cure many human infections, especially wound infections, which are mainly caused by Staphylococcus bacteria.

Fleming's discovery about the killing action of Penicillium mold led to the first commercially produced antibiotic called penicillin. Following the successful production of penicillin, many other antibiotics have been developed and used to cure numerous bacterial infections. During the Civil War, over 50 percent of the surgery patients died from bacterial wound infections. Due to the use of antibiotics and various other improved techniques, it is extremely rare for a modern surgery patient to die because of a surgery wound infection. How many people do you know who might not be alive today without antibiotics?

© Mark Twain Media, Inc., Publishers

MY CONCLUSIONS

The main **conclusion** for any science investigation is a statement that either supports or rejects the hypothesis of the project. In the example on page 32, Jim's hypothesis was that wind will increase the rate of water loss (transpiration) from the leaves of Zebrina plants. As shown below, the bar graph for Jim's two trials provides solid support for his hypothesis.

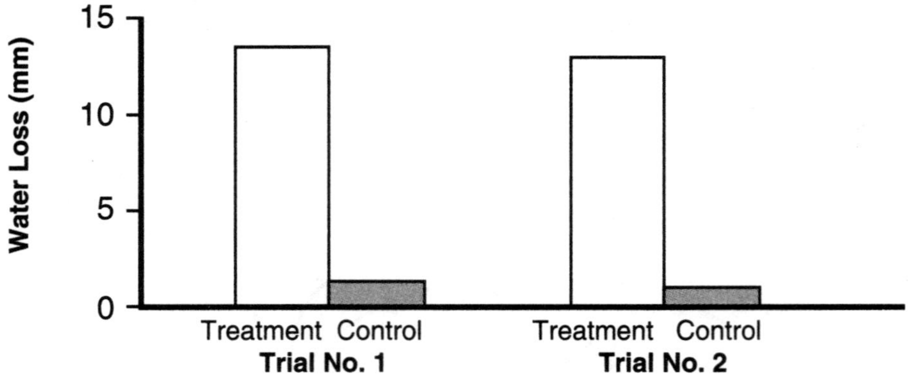

Jim's main conclusion, therefore, can be that *wind causes increased water loss (transpiration) from Zebrina leaves—at least under the conditions of the experiment*. Jim may have gotten different results if the time of treatment had been six hours rather than one hour or if the temperature had been 100° Fahrenheit rather than room temperature (72° F). You must always be very careful not to **over-conclude**. If Jim would state in his conclusions that wind always causes increased transpiration, then he would be over-concluding!

A successful science investigation project will usually suggest further experiments. Jim's results might cause him to wonder about how the speed of the wind would affect the rate of transpiration in Zebrina. Or he might wonder if wind would affect *other kinds of plants* in the same way that it affects Zebrina. These ideas for further investigation could be discussed in the "conclusion" section of Jim's project report.

No investigation is perfect. In an experiment, there are always some problems with control, gathering data, and other aspects of the scientific method. In Jim's experiment, ensuring that there was no difference in the average leaf size between the treatment and control groups is a control problem that could also be discussed in the conclusion section of Jim's project report. Jim can review how he measured the size (length x width) of all leaves on a sprig and how he assigned sprigs to control and treatment groups so that the total leaf size (surface area) in each group was very similar. Jim should also explain how he switched the treatment and control groups for the second trial of the experiment to help control for any differences in leaf size.

Would Jim's experiment have been successful if his results caused him to *reject his hypothesis*? YES. Any well-designed and carefully-performed investigation is successful regardless of the results obtained. In fact, it is always wise to remember that we can never prove anything in science with 100 percent certainty. We can only support or reject hypotheses. Rejecting the stated hypothesis of an investigation means that you have discovered something about your hypothesis by eliminating one possible answer. All good science investigations improve our understanding of the material world.

© Mark Twain Media, Inc., Publishers

SHARING MY RESULTS

The Final Written Paper

The final step for any successful science investigation project is sharing your results with others. In writing your final paper, you should use the same general format that a scientist would use reporting her or his results in a scientific journal. This paper should include a title and the following sections: introduction, methods and materials, results, conclusions and discussion, and citations and references. Some ideas about the content of each section of the final paper have been discussed earlier in this book, but it is very important that the final written paper be very concise—with no more information than is necessary.

The title in a scientific paper generally describes what an experiment is about. "The Effect of A On B" is a common form for the title of a scientific paper. Fill in "A" with the *independent variable* and "B" with the *dependent variable* of the experiment, and you have an acceptable title. In Jim's experiment (described on pages 36-38), an acceptable title would be "The Effect of Wind On Transpiration In Zebrina Plants." This title is very descriptive but it may seem rather dry. Today, many middle school and high school project titles are designed to catch the attention of the reader (or science fair or science academy judge) rather than to be very descriptive. (The project is fully described in the abstract and body of the final written paper.) If you look at the titles on pages 15-17, you see both descriptive titles for science projects and "catchy," non-descriptive project titles.

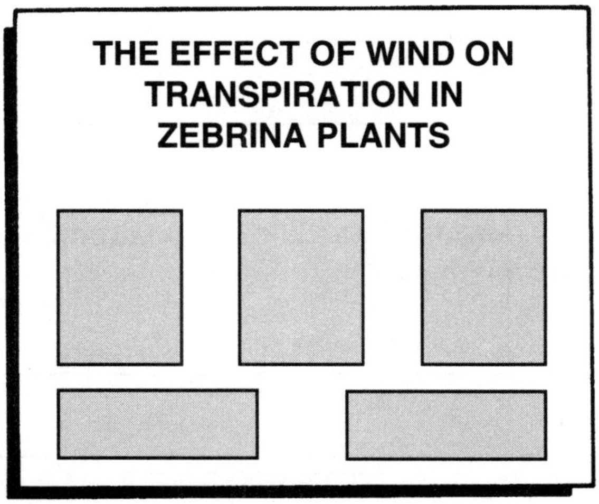

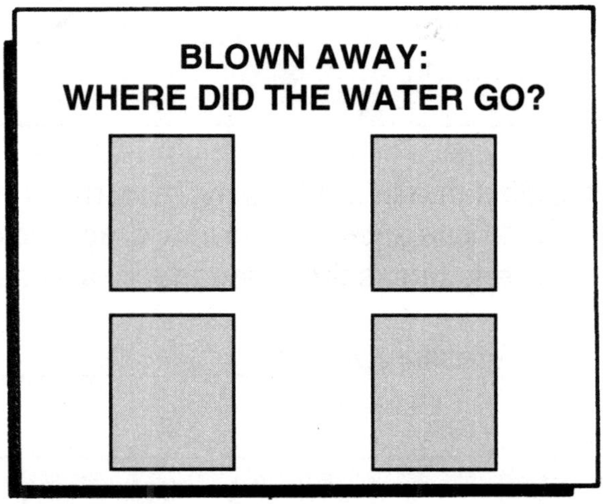

© Mark Twain Media, Inc., Publishers

Name ______________________________ Date ____________________

Student Exercise

1. Use the list of science project titles on pages 15-17 to find five descriptive titles that follow the form "The Effect of A on B," where "A" is the independent variable and "B" is the dependent variable. List these titles below.

   a. ______________________________

   b. ______________________________

   c. ______________________________

   d. ______________________________

   e. ______________________________

2. Use the list of science project titles on pages 15-17 to find five "catchy," non-descriptive titles. List these titles below.

   a. ______________________________

   b. ______________________________

   c. ______________________________

   d. ______________________________

   e. ______________________________

3. Select one of the "catchy," non-descriptive titles from your list of answers for question two above and use your imagination to change it to a descriptive title. (You will probably have to guess the independent and dependent variables.)

   Non-descriptive title: ______________________________

   ____

   Descriptive title: ______________________________

   ____

   ____

© Mark Twain Media, Inc., Publishers

Sections of the Paper

The topic question to be explored in a science investigation and the hypothesis are usually explained in the **introduction** of the final paper. A discussion of what is known about the topic is also typically included under the introduction. This discussion helps the reader to understand the scope and purpose of the project.

The **methods and materials** section describes the essential steps or procedures that were completed during the project. The equipment and other supplies utilized are also described in this section.

In the **results** section of the paper, the data from the investigation are summarized. Tables and graphs are often used for this purpose. (See pages 39-52 for a discussion of tables and graphs in this book.)

The main emphasis of the **conclusions and discussion** section of the final written paper is an explanation of why you accept or reject your hypothesis. (See pages 54-55 of this book for more information about conclusions in a scientific paper.) However, you should also discuss possible sources of error in your investigation and any ideas that could improve or expand your science project.

Separate listings for citations and references should follow the conclusions and discussion section of your paper. A **citation list** should be given first, followed by a separate **reference list**. Please remember that *citations* mean all of those sources that you made direct reference to in your paper. A method of citing articles and books that is used in scientific journals is given below.

Citation in the Body of the Paper:

> *It has been shown in at least one study (Smith and Jones, 1979) that vitamin A causes some increases in regeneration in starfish.*

Citation (or Reference) Listing for an Article:

> *Smith, P.C. and J.R. Jones. 1979. A Study of Starfish Regeneration. The Journal of Invertebrate Reproductive Biology. Vol. 6, No. 3, P. 14-19.*

Citation (or Reference) Listing for a Book:

> *Smith, P.C. and J.R. Jones. 1979. A Study of Starfish Regeneration. 1st Ed. University of Iowa Press, Iowa City, Iowa.*

References are sources that were read and that may have influenced your investigation but were not cited in the body of the paper. The reference list follows the exact same format as the citation list. The citation and reference lists should each be given in alphabetical order by author.

Science teachers often require a minimum number of citations and references. However, most teachers realize that there may be great differences in the number of library sources available for the various project topics in a science class.

© Mark Twain Media, Inc., Publishers

Abstracts are always required for studies that are published in scientific journals, and they are also used in any middle school and high school science fair displays and science academy papers. An **abstract** is a short (often less than 100 words) summary of the entire science project. The project title and abstract together should provide a good overview of your science investigation.

At the very end of your final written paper, there should be a brief **acknowledgments** section where you thank all those individuals who provided you with help during your science project.

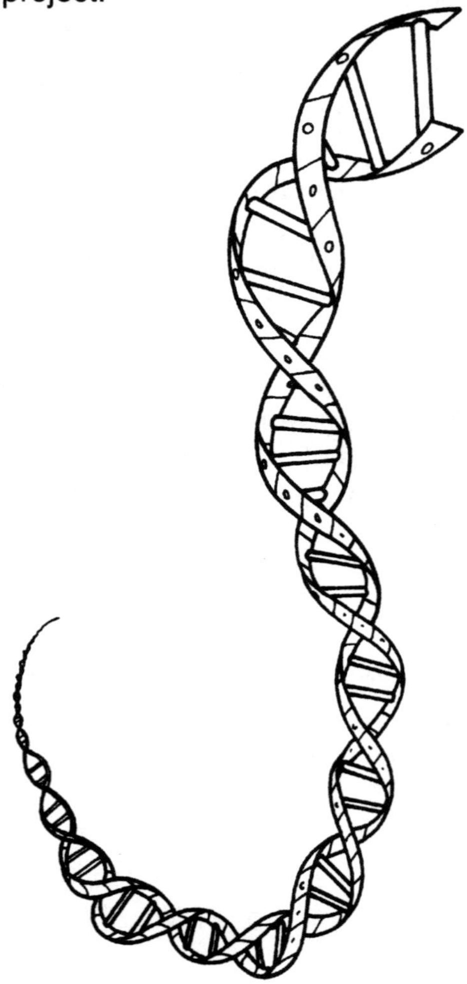

One of the great advances of modern biology has been the discovery of the structure and function of the DNA molecule. This discovery has led to the use of DNA fingerprinting in criminal investigations and to advances in biotechnology, such as the production of new kinds of bacteria that are capable of growing large quantities of needed substances like insulin and human growth hormone. The cloning of higher animals, such as the now famous ewe named "Dolly," is also based on the understanding of DNA.

The article that first explained the structure of the DNA molecule was published by James Watson and Francis Crick in the scientific journal, *Nature*, in 1953. The entire article, titled "Molecular Structure of Nucleic Acids," was approximately one full page of the journal. Crick and Watson certainly understood the value of **concise reporting of results**! Crick and Watson received the Nobel Prize in Biochemistry for their work.

The Final Display Board

Many science fair **display boards** use sturdy pieces of cardboard taped together to form a freestanding display as shown below. Standard-size dimensions are given.

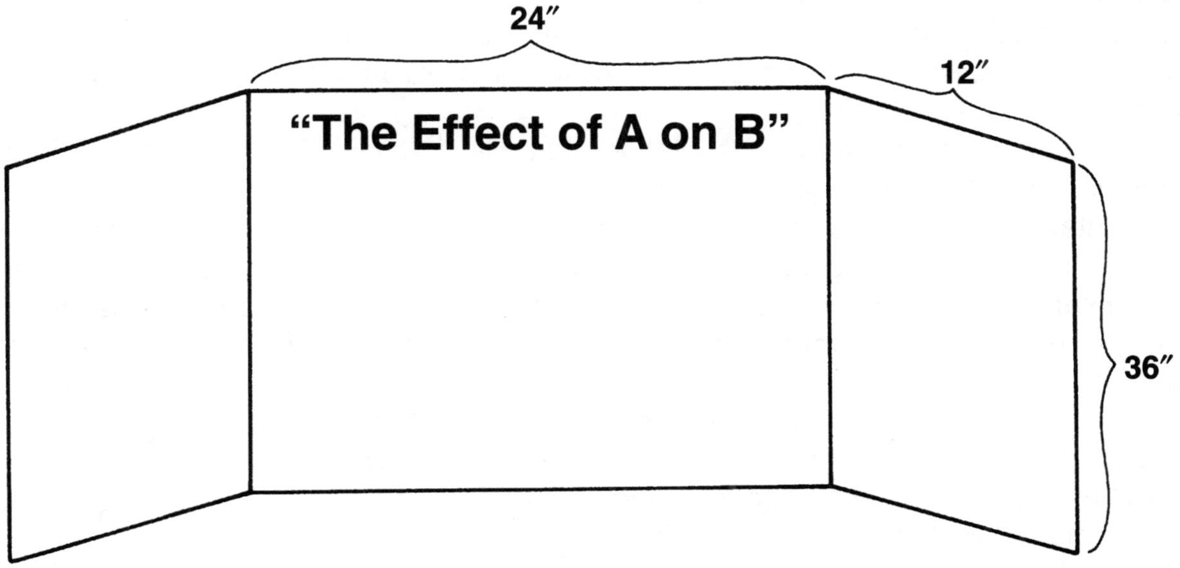

Many school supply stores now sell standard one-piece project boards. Project boards can also be constructed from plywood, pressed board, or foam-core poster board pieces hinged together. These types of board are often covered with felt so that various sized poster board pieces can be attached with Velcro.

Even though the standard display board size given above is used for most middle school and high school science fairs, there are some differences in size requirements. Also, some science fairs permit some of the project equipment and a copy of the final written paper to be part of the project display. Therefore, it is essential to **know the requirements of the particular fair that your project will enter**.

© Mark Twain Media, Inc., Publishers

The Final Oral Report

Formal oral reports of science investigations are given by students who enter regional and state academies of science competitions. (Science academies are usually organized at the state level, but there may be regional science competitions that involve several counties.) These competitions are typically held on college campuses. Until recent years, science academies only included research presentations from high school, college, and adult scientists, but now middle school students are able to compete in many of these academies.

During the academy presentation, you will be allowed a set amount of time to summarize and discuss your science project. In middle school academies, time for presentation usually varies between eight and 12 minutes, with at least two to three minutes of questions from the judges. A warning sign is usually shown to the student presenter with one minute of time remaining in the formal presentation, but no extra time is usually given. This is why **you must practice your oral report many times before you present it at the academy**.

Do not simply read your final written paper during your oral report. A good way to do your oral report is to work from an outline of your talk on note cards. Also, the use of audiovisuals, such as overhead transparencies, 2″ x 2″ slides, and computer presentation programs like PowerPoint, can greatly enhance your talk.

After the oral presentation, there is usually time provided for questions from the judges. Below are some examples of good questions that may be asked by the judges.

What got you interested in your topic?
How did you arrive at your hypothesis?
Describe your controls.
What is your independent variable?
What is your dependent variable?
Could your experiment be duplicated by others?
How many times did you repeat your experiment?
How could you improve upon your investigation?

Practical application?

These are also some of the very same questions that are asked by science fair judges when there is time to talk with students at their display boards. Any good science competition will provide student participants time to **communicate** about their projects.

Do not forget to send letters or cards thanking all of the people who helped you with your project.

A TIME LINE AND SOME STANDARD LETTERS AND FORMS

Middle school and high school science competitions usually take place during the spring semester. This provides ample time throughout the school year for students to design and complete science investigation projects. Even though the students are the owners of their investigations and should have the final say about everything that goes into them, teachers and parents are always needed to provide guidance, encouragement, and expertise.

In order to ensure that students complete their science investigations and are ready for the spring science fair or science academies, the science teacher should develop a **project time line** that includes all of the **student deadlines**, as well as mailings to parents and other adult participants. The following time line is an example for a middle grade science teacher (let's call him Mr. Adams) who is sponsoring students for a **regional science fair** and a **regional science academy**. Both of these events are to take place in April. Mr. Adams plans to coordinate and direct a science fair for the students in his class on the first Tuesday in March, from 7 to 9 p.m. This **class science fair** will be held in the school cafeteria and will be open to the public.

In an ordinary school year, students should be working with science investigation topic questions and hypotheses early in September. This early stage is also a good time for Mr. Adams to **let the school principal and school librarian know that students will be doing science investigation projects** throughout much of the school year and that they will be presenting them in the spring. A **letter to the parents or guardians** of all science investigation students will be sent by Mr. Adams at this time. This letter will explain some of the benefits that students gain from doing science investigations and will also describe what is expected of the students and parents as projects are developed and presented.

A sample letter to parents is given on the next page.

© Mark Twain Media, Inc., Publishers

Sample Letter to Parents

September 14, 1999

Mr. and Mrs. William Jacobson
2322 Park Drive
Maysville, KS

Dear Mr. and Mrs. Jacobson,

This year, all of my seventh grade science students will be developing science investigations projects. They will be working on these projects from early September through most of April. The students will first present the results of their science projects at a class science fair called "Having Fun with Science Night" on March 7 and also at a regional science fair and a regional science academy in April. You are cordially invited to all of these events. (More information will be provided later.)

Your daughter, Lisa, should have complete ownership of her project and make all final decisions about what the project will include. However, your support and encouragement will be very helpful to Lisa. I know that Lisa can count on both of you.

I have included a copy of my science investigation project time line for this school year so that you know about all of the deadlines that Lisa must meet in completing her science project. Please feel free to call me at 555-3987 if you have any questions. The best time to reach me is during my planning period, 1:45–2:30, Monday through Friday.

Sincerely,

Mr. Arnold Adams
Science Teacher

Since Mr. Adams is planning a science fair for the spring, he needs to **reserve a room (or rooms)** for the event. He intends to reserve either the gymnasium or the cafeteria, because each has enough room to house tables for freestanding student display boards.

By the end of September, Mr. Adams intends to have taken his science class on **field trips to visit both the school and local public libraries**. By the end of the second week of October, all of Mr. Adams' students should have explored the library, the World Wide Web, and other sources of information about their topics. **Project proposals** that include the *topic question*, *hypothesis*, *feasibility study*, and *experimental design* should be submitted to Mr. Adams by the end of October. Mr. Adams plans to approve all project proposals (with possible suggested modifications) by the end of the first week of November. Through the rest of November, December, January, and February, Mr. Adams' students will be working individually on their science investigations projects.

Mr. Adams will require a *one-page project report* from each of his students at the end of the second week of December and at the end of January. These project reports will include a description of what steps have been accomplished and what data has been gathered. Mr. Adams will meet with each student individually to discuss the project report.

© Mark Twain Media, Inc., Publishers

Sample Science Fair Judge Recruitment Letter

Mr. Adams intends to contact potential judges for his class science fair, which is scheduled for the first Tuesday evening in March. Mr. Adams has decided that the class science fair will be noncompetitive and that the judges will evaluate the student science projects without selecting any winners or ranking the projects. According to Mr. Adams, the class science fair will be called "Science Fun Night," and it will be an opportunity for the students to share their projects with one another and learn from the judges. There will still be some time for the students to make some final improvements in getting ready for the regional science fair and science academy. Both of these events will be competitive.

An example of a science fair judge recruitment letter is given below.

January 14, 1999

Dr. Patricia Marlow
Department of Physics
Midwestern University
Oakbrook, Kansas

Dear Dr. Marlow,

On March 7, students in my seventh grade science class will be presenting the results of their science investigation projects in a science fair. This event will take place in the school cafeteria from 7 to 9 p.m. Between 8:30 and 9 p.m., there will be an awards ceremony where all participants will be recognized for completing work on their projects.

All of my students intend to present their projects in both the regional science fair and the regional science academy later in April. Therefore, it is very important that they get an opportunity to practice answering questions posed by qualified judges. It is also important that the students receive input about their projects so that they can make some final improvements before the regional competitions. I would like to know if you would be available and willing to serve as a science fair category judge for our class fair. I realize that you have a very demanding schedule, but I can assure you that your expertise and experience would be of great benefit to the students, and we would greatly appreciate your service. I will call you in approximately two weeks to receive your answer. If you would like to call me before then, my number at school is 555-3987.

Thank you for your consideration.

Sincerely,

Mr. Arnold Adams
Science Teacher
Maysville Middle School
Maysville, KS

Sample Science Fair Judging Form

An example of a typical middle school science fair judging form is given below. Except for question seven, all of the questions could also be used to evaluate the oral presentations and final written papers in a science academy.

Regional Science Fair

Category: ______________________ Project Number: __________

	LOW	AVERAGE	HIGH
1. Originality of the topic Is the student aware of other work in the area?	1 2 3	4 5 6 7	8 9 10
2. Problem is well defined Is it testable, reproducible, falsifiable?	1 2 3	4 5 6 7	8 9 10
3. Research paper covers previous work Are there enough listings in the bibliography?	1 2 3	4 5 6 7	8 9 10
4. Methods are clearly described Are materials listed and explained?	1 2 3	4 5 6 7	8 9 10
5. The data is displayed Has the experiment been repeated enough?	1 2 3	4 5 6 7	8 9 10
6. Conclusion of results Does data support the conclusions?	1 2 3	4 5 6 7	8 9 10
7. Visual appeal of display Does the display show creativity?	1 2 3	4 5 6 7	8 9 10
8. Oral presentation Can the student answer questions, as well as speak from a prepared dialogue?	1 2 3	4 5 6 7	8 9 10
9. Investigative nature of the project Are there controls and manipulated variables?	1 2 3	4 5 6 7	8 9 10
10. Student's knowledge of topic Does the student understand the topic and project conclusions?	1 2 3	4 5 6 7	8 9 10

RIBBON RECOMMENDATION:
Blue "Superior" 70–100 points
Red "Excellent" 40–69 points
White "Good" 10–39 points
* A purple ribbon may also be awarded for the best project in each category.

TOTAL POINTS = __________

List of Science Fair Categories

A competitive local or class science fair has the advantage of being able to select only the top winners to participate in regional fairs and academies. Some regional science fairs and academies actually require this type of selection process.

Science fair and science academy projects are typically grouped into categories. Listed below are the current categories from the *International Rules for Precollege Science Research: Guidelines for Science Fairs*, sponsored by Science Service, Inc., Washington, D.C. These same categories are often used in many other science fairs and academies.

BEHAVIORAL AND SOCIAL SCIENCES
BIOCHEMISTRY
BOTANY
CHEMISTRY
COMPUTER SCIENCE
EARTH AND SPACE SCIENCE
ENGINEERING
ENVIRONMENTAL SCIENCES
MATHEMATICS
MEDICINE AND HEALTH
MICROBIOLOGY
PHYSICS
ZOOLOGY
TEAM PROJECTS

In each category, place awards are usually given for first, second, and third places. Any number of honorable mention awards may also be given. Ribbon awards are generally blue for the highest, then red, followed by white. A purple ribbon is often awarded for the best overall project in a category. In addition, one of two grand winners for the entire fair are sometimes selected and presented with special trophies.

Many middle grade and high school science competitions, including the International Science and Engineering Fair (ISEF) that is sponsored by Science Service, Inc., have special categories for team projects, as well as individual categories. As shown in the category list above, the ISEF uses a separate category called "TEAM PROJECTS" that includes team projects from any subject area. As an alternative, team projects can be considered separately within each subject category. Team projects with two or three students working together encourage cooperation and can make the project an easier task for the students. If you examine **scientific journal articles**, you will find that **most have multiple authors who have worked together on the investigation described in the article**.

© Mark Twain Media, Inc., Publishers

Science Investigations Time Line

Mr. Adams' time line for his science investigations program is given below.

Completed By	
September 14	• Students working on topic questions and hypotheses • School principal and school librarian informed about science investigations projects • Letter describing science investigations program sent to parents or guardians
September 30	• Students working on use of library and Internet sources • Class field trips to school and public libraries • Class science fair room (or rooms) reserved
October 7	• Students working on experimental design of science investigations
October 21	• Student project proposals due
November 7	• Project proposals approved and returned to students (individual meetings)
December 14	• First progress report due from each student (individual meetings)
January 30	• Second progress report due from each student • Judges for class science fair secured (formal letters sent)
February 14	• Second letter sent to parents and judges, reminding them about class science fair • Obtain ribbons, prizes, or any other supplies or materials needed for the class science fair
March 7	• Class science fair ("Science Fun Day")
March 30	• Students improving projects and getting ready for regional science events
April 7	• Regional science fair
April 14	• Regional science academy
April 30	• Evaluate the success of the science events • Send "thank you" letters to parents, judges, and all others who helped with the student projects and science events

© Mark Twain Media, Inc., Publishers

Focus Group Discussion Guidelines

The final task for Mr. Adams in his school science investigations program is to find out how successful the program was and what measures can be instituted to improve the program in the future. To accomplish this, Mr. Adams has decided to organize a **focus group discussion** at the school during one of his planning periods. He will invite the school principal, the school librarian, one of the class science fair judges, one parent (or guardian), one student, and one teacher who will take detailed notes on the discussion. Mr. Adams will lead the discussion based on the following general questions.

1. Did the science investigations program help the students to better understand science and its importance in today's world?
2. Did the science investigations program help the students to learn new skills and to perform at higher cognitive levels such as analysis, synthesis, and evaluation?
3. What were some of the negative results of the science investigations program?
4. Did the science investigations program motivate students to learn more about science?
5. Were there any unexpected benefits from the science investigations program?
6. Were the benefits derived from the science investigations program worth the time and effort put into it?
7. What are some of the ways that the science investigations program can be improved in future years?

In order to gather more meaningful input about the science investigations program, Mr. Adams may ask these same questions of all parents and guardians of science investigations students on a questionnaire.

© Mark Twain Media, Inc., Publishers

Name ______________________________ Date ________________

Sample Student Science Investigations Evaluation Form

In addition to the focus group and parent/guardian questionnaire, Mr. Adams will require each science investigations student to complete the following questionnaire.

STUDENT SCIENCE INVESTIGATIONS EVALUATION

Name: ______________________________ Date: ________________

Project Title: ____

____

Respond to each of the following questions using the following scale: 1—strongly agree, 2—agree, 3—neutral, 4—disagree, 5—strongly disagree. Circle one number for each statement.

1. I learned quite a bit about how science is actually done by doing my science investigation project.

   1 2 3 4 5

2. I enjoyed doing my science project.

   1 2 3 4 5

3. I am looking forward to doing another science investigation project.

   1 2 3 4 5

4. I felt a sense of accomplishment by completing my science project.

   1 2 3 4 5

5. I spent a great amount of time in completing my science investigation project.

   1 2 3 4 5

6. The time that I spent on my science investigation project was well worth it.

   1 2 3 4 5

Name ______________________________ Date ______________

Rank the following 10 science investigation project steps from most difficult (1) to least difficult (10).

_____ Selecting a Topic

_____ Developing a Hypothesis

_____ Doing the Library and Internet Search

_____ Designing the Experiment

_____ Collecting the Data

_____ Organizing the Data into Tables, Graphs, etc.

_____ Drawing Conclusions

_____ Writing the Final Paper

_____ Putting Together the Display Board

_____ Giving the Oral Report

Describe two unexpected problems that occurred during your science investigation project.

____

____

____

____

____

Describe two ways to improve your project.

____

____

____

____

____

© Mark Twain Media, Inc., Publishers

22 SCIENCE TEACHER "DO'S"

1. Establish a time line early in the school year and stick to it.

2. Encourage your students to pick a science investigation topic in which they are interested.

3. Require an approved project proposal before any student begins collecting data.

4. Schedule time for individual discussions with each science investigations student throughout his or her project.

5. Make certain that all students observe standard safety procedures when working on their projects. (Remember that close adult supervision is necessary for many procedures.)

6. Take your students to the school and public libraries, and be certain that they understand how to find information about their topics. (Stress the importance of a thorough library search as part of their investigation, and help them find information on the World Wide Web.)

7. Encourage your students to do science investigation projects that explore new ideas that are not standard textbook demonstrations.

8. Be a facilitator—offer helpful suggestions and encouragement, but do not do the projects for your students. (The students should have ownership of their projects.)

9. Stress the noncompetitive aspects of science fairs and science academies. (Remind your students that everyone who completes and presents a science investigation project is a winner because they have found out something new and have shared it with others.)

10. Remind your students that it is natural for them to feel "lost" at certain times during their science investigation projects. (Science investigations involve a search for something new. Doing science has been jokingly described as "what you do when you do not know what you are doing." There is some truth to this statement.)

11. Remind your students that projects do not have to be complicated to be good.

12. Encourage your students to test one variable at a time in their experiments.

13. Encourage your students to do enough trials in their experiments.

14. Encourage your students to control as many variables as possible in their experiments.

© Mark Twain Media, Inc., Publishers

15. Encourage your students to use sufficiently large samples in their experiments.

16. Encourage your students to use plants or invertebrate animals rather than vertebrate animals in their laboratory experiments. (Since there are many strict laws that regulate the use of vertebrate animals, it is recommended that students using vertebrates follow the current *International Rules for Precollege Students Research: Guideline for Science Fairs*, Science Service, Inc., Washington, D.C.)

17. Remind your students that a science investigation that rejects the original hypothesis is a perfectly valid investigation. (Also remind them that science can never really prove a hypothesis with certainty.)

18. Encourage your students to keep good records of all data.

19. Encourage your students to expect the unexpected.

20. Encourage your students to represent their data in summary form by using tables and graphs.

21. Remind your students to draw only conclusions that are based on the data in the results of the investigation.

22. Encourage your students to formally thank all individuals who helped them with their science investigation project.

LIST OF SCHOOL SCIENCE SUPPLY COMPANIES

Wards
P.O. Box 92912
Rochester, NY 14629-9012
www.wardsci.com

Carolina Biological Supply Co.
P.O. Box 6010
Burlington, NC 27216-6010
carolina.carolina.com

Nasco Science
901 Janesville Ave.
Fort Atkinson, WI 53538-0901
or
4825 Stoddard Rd.
Modesto, CA 95356-9318
info@nascofa.com

Edmund Scientific Co.
Science Division
101 East Gloucester Pike
Barrington, NJ 08007-1380
scientifics@edsci.com

Fisher Scientific
U.S. Headquarters
2000 Park Lane Dr.
Pittsburgh, PA 15275-9952
www.fishersci.com

Nebraska Scientific
3823 Leavenworth Street
Omaha, NE 68105-1180
nescientif@aol.com

Sargent Welch
P.O. Box 5229
Buffalo Grove, IL 60089-5229
www.sargentwelch.com

Sigma Chemical
P.O. 14508
St. Louis, MO 63178
www.sigma.sial.com

© Mark Twain Media, Inc., Publishers

SOME PARENT "DO'S"

1. Let your child know that you understand the importance of science in our world. (Attempt to discuss current science events from television and newspapers with your child.)

2. Show support for the idea of student science investigation projects as a way to learn more about the process of science. (Educators overwhelmingly support this type of investigative activity as a way to improve critical thinking skills in students.)

3. Encourage your child to pick a science investigation topic that she or he is truly interested in. It is okay for parents to help out with some early topic suggestions, especially if the suggestions lead to other topic ideas by your child. (It is very important that your child assumes ownership of her or his project early in the investigative process.)

4. Help your child to remember the science investigation deadlines that are set by the science teacher. (Notes on a household calendar may serve this purpose.)

5. Volunteer your time and expertise to help your child's science teacher with the science investigation program. Science fairs and academies cannot succeed without a certain amount of parent involvement. Parents can help with registration, fund-raising, the awards ceremony, building displays, facility arrangements, and as consultants for the student participants. (However, be very careful not to participate in activities that might prove to be a conflict of interests, such as judging a category in which your child's project is entered.)

6. Caution your child to learn and observe safety procedures during all science investigation project work. (Attempt to anticipate any possible safety problems that might occur with your child's project before she or he begins work.)

7. Be willing to take your child (and possibly some classmates) to a local library to find information about her or his science investigation topic. (If possible, help your child find information on the World Wide Web.)

8. Let your child know that you are willing to help her or him with any phase of the science investigation project, but be certain that your child does the actual work and makes all of the final decisions about the project. (Remember that student ownership of the project is essential for a successful science investigation experience.)

9. Stress the value of sharing what is learned from a science investigation in a noncompetitive way. (Remind your child that everyone who completes and presents a science investigation project is a winner because they have discovered new information and have shared it with others.)

© Mark Twain Media, Inc., Publishers

10. Remind your child that it is perfectly okay to feel "lost" at certain times during the science investigation process because she or he is dealing with something new and/or different. (Stress the importance of rethinking procedures, gathering new information from the literature, and seeking direct help from the science teacher and other "experts" in order to overcome certain problems during the science investigation process.)

11. Encourage your child to treat all experimental organisms with respect and in a humane manner. The science teacher should provide your child with specific guidelines about the humane treatment of experimental organisms. (Important lessons about "respect for life" can be accomplished during a student science investigation.)

12. Remind your child that a project does not have to be complicated to be a good one.

13. Encourage your child to keep good records of all data.

14. Encourage your child to formally thank all individuals who helped her or him with the science investigation project—especially the classroom science teacher. A letter of thanks to your child's science teacher (with a CC to the school principal) for all of the extra time, effort, and ingenuity that was required to coordinate the science investigation program is appropriate.

© Mark Twain Media, Inc., Publishers

Name ______________________________ Date ______________

SCIENCE INVESTIGATIONS REVIEW: Word Search

Find the words listed below and circle them in the word search puzzle. Words may be written forwards or backwards and horizontally, vertically, or diagonally.

Q	L	C	O	N	C	L	U	S	I	O	N	S	M	T	L	I	F	E	C
W	S	D	S	E	S	E	H	T	O	P	Y	H	S	R	V	C	P	S	U
R	E	P	E	A	T	A	B	I	L	I	T	Y	Q	K	Q	E	H	S	V
T	B	A	F	F	X	J	D	V	A	R	I	A	B	L	E	R	Y	J	A
P	H	A	K	E	E	X	P	E	R	I	M	E	N	T	A	L	S	D	U
H	D	K	X	D	I	S	P	L	A	Y	V	H	P	A	R	G	I	C	C
L	L	M	S	I	P	H	L	F	D	Y	E	L	B	A	T	G	C	O	O
A	S	A	I	B	N	E	V	I	T	P	I	R	C	S	E	D	S	L	R
S	I	G	N	C	R	V	N	O	I	T	A	V	R	E	S	B	O	M	R
O	N	V	E	R	R	U	E	W	C	O	M	M	U	N	I	C	A	T	E
P	J	M	U	O	U	O	V	S	S	R	N	J	L	Q	F	U	B	C	L
O	F	J	I	Q	L	O	B	K	T	Y	I	K	E	T	G	O	Y	H	A
R	A	T	A	D	V	O	J	I	N	I	W	A	D	S	T	I	N	E	T
P	W	W	C	S	Q	U	G	X	O	W	G	I	F	A	M	S	E	M	I
M	F	T	R	I	A	L	S	Y	T	L	J	A	N	S	C	T	I	I	O
F	K	Y	M	N	L	R	N	D	N	Q	O	Y	T	I	S	R	I	S	N
V	G	S	L	C	L	O	R	T	N	O	C	G	E	I	C	S	I	T	F
N	S	E	C	N	E	R	E	F	E	R	S	N	Y	J	O	C	S	R	Y
Q	N	I	S	N	O	I	T	A	T	I	C	K	D	H	G	N	H	Y	I
L	A	C	I	S	Y	H	P	J	E	E	M	A	T	E	R	I	A	L	S

WORD LIST

BOTANY
CONCLUSIONS
DESCRIPTIVE
GEOLOGY
JOURNAL
OBSERVATION
REFERENCES
TRIALS
CHEMISTRY
CONTROL
DISPLAY
GRAPH
LIFE
PHYSICAL
REPEATABILITY
VARIABLE
CITATIONS
CORRELATION
EXPERIMENTAL
HYPOTHESES
MATERIAL
PHYSICS
SCIENCE
COMMUNICATE
DATA
FAIR
INVESTIGATION
MICROBIOLOGY
PROPOSAL
TABLE

© Mark Twain Media, Inc., Publishers

Name ________________________________ Date ________________

SCIENCE INVESTIGATIONS REVIEW: Crossword

Use the clues below to complete the crossword puzzle.

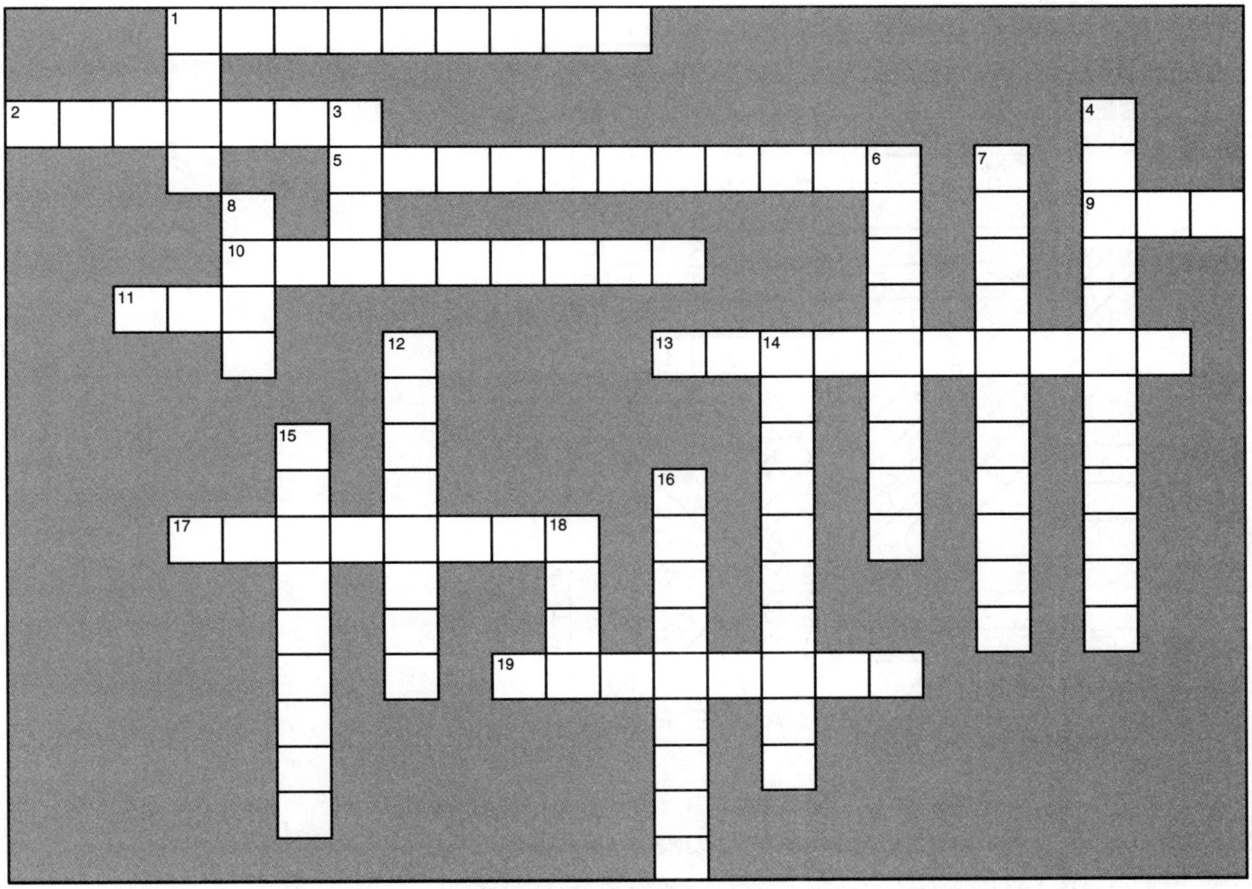

ACROSS

1. The variable that is measured during a scientific experiment
2. The group in an experiment that does not receive a treatment
5. The variable that an investigator changes during an experiment
9. A _____ chart shows the percentages of all possible observations.
10. A way to get a single quantity that represents a group of several quantities
11. A graph that can easily summarize and compare several totals
13. A topic question that has been reworded into a form that can be tested
17. Science can only deal with the _____ world.
19. The file may contain newspaper and magazine articles on a particular subject.

DOWN

1. Results collected during your investigation
3. All of the sciences can be divided into the physical and the _____ sciences.
4. These studies consider all variables in order to establish the cause of observations.
6. The group in an experiment that receives the experimental condition
7. Shows a connection among observations without determining causes
8. All of the books in the library are recorded in the _____ catalog.
12. A short summary of the entire science project
14. In a library, magazine articles can be found in the *Guide to _____ Literature.*
15. All of the sources that are referred to directly in the final paper are called _____.
16. Good science investigation _______ can be answered by selecting one answer from a few possible choices.
18. A graph that shows how two variables change in relation to one another

© Mark Twain Media, Inc., Publishers

SCIENCE INVESTIGATIONS REVIEW: Answer Keys

Science Investigations Word Search (page 74)

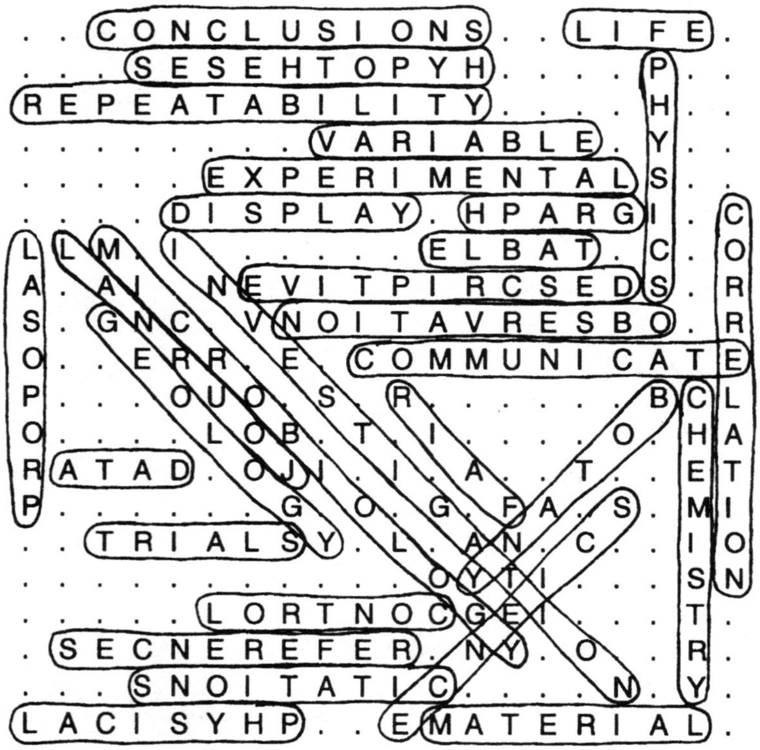

Science Investigations Crossword Puzzle (page 75)

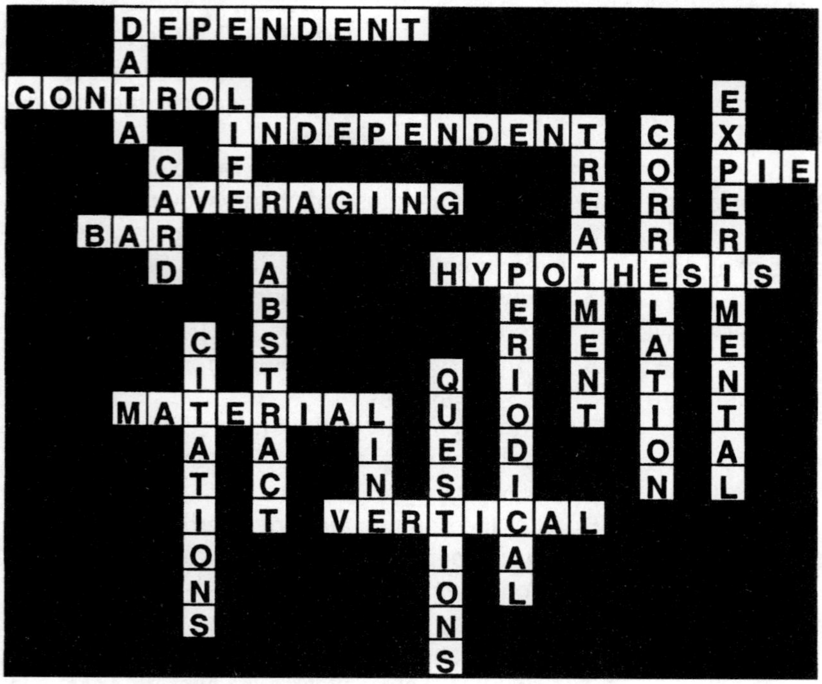

© Mark Twain Media, Inc., Publishers